Matter, Ether, and Motion

THE FACTORS AND RELATIONS OF PHYSICAL SCIENCE

A. E. DOLBEAR Ph.D.

PROFESSOR OF PHYSICS TUFTS COLLEGE

AUTHOR OF "THE TELEPHONE"

"THE ART OF PROJECTING" ETC.

Published by

Hawk Press
4836/24, Ansari Road, Daryaganj
New Delhi – 110 002
Phones : 9643330713, 011-23278618, 011-35676207
E-mail : thehawkpress@gmail.com
www.thehawkpress.com

ISBN : 978-93-93971-76-0

PREFACE TO THE SECOND EDITION

The issue of a new edition of this book gives me an opportunity to make some needed corrections, and enlarge it by the addition of three new chapters, which I hope will make it more useful to such as have a taste for fundamental physical problems. The first of these, Properties of Matter as Modes of Motion, presents the evidence that all the characteristic properties of matter are due to energy embodied in various forms of motion. The second, on The Implications of Physical Phenomena, points out what assumptions are made in explaining phenomena. It is the substance of a series of articles published in the *Psychical Review* in 1892 and 1893. The third, on The Relations between Physical and Psychical Phenomena, was read as a paper before the Psychical Congress at the World's Fair in August, 1893.

Judging from some of the comments made about my statements as to Modern Geometry on page 67, and as to Vital Force, p. 336, I have thought it would be useful to some to see corroboratory statements; and I have therefore added, in an appendix, a few pages of quotations from some of the most eminent mathematicians and biologists on these subjects, and from them one may judge whether or not my statements are correct.

As the work is a treatise on Physics, there is no special reason for going beyond it; but if this presentation of the subject is any approach to the truth, there is an important conclusion to be drawn from it. If the ether be the homogeneous and uniform medium it is believed with reason to be, then, in the absence of what we call matter, no physical change which we call a phenomenon could possibly arise in it; for every such phenomenon is a product, and in the absence of one of the essential factors, viz., matter, it could not be. If matter itself be a form of motion of the ether, the ether must have existed prior to matter; also, if the atom be a form of energy, then must energy have existed before matter existed. Hence there must have been some other agency radically different from any physical energy we know, and independent of everything we know, which was capable of producing orderly physical phenomena, by acting upon the ether; for a homogeneous medium could not originate it. Some philosophers call this antecedent power The Unknowable; others call it God. If energy *as we know it* implies antecedent energy as we do not know it, so, likewise, mind as we know it implies antecedent mind under totally different conditions from those in which we find it embodied.

In whatever direction one pursues physical science, he is at last confronted with a physical phenomenon with a superphysical antecedent where all physical methods of investigation are impotent. Such considerations raise the theistic hypothesis of creation to the rank of such physical theories as the nebula theory of the origin of the solar system, and the undulatory theory of light.

PREFACE

WITHIN the past fifty years the advance in physical knowledge has not only been rapid, but it has been well-nigh revolutionary. Not that knowledge that was felt to be well grounded before has been set aside,—for it has not been,—but the fundamental principles of natural philosophy that were applied by Sir Isaac Newton and others to masses of visible magnitude have been applied to molecules; and it has thus been discovered that all kinds of phenomena are subject to the same mechanical laws. It was thought before that physics embraced several distinct provinces of knowledge which were not necessarily related to each other, such as mechanics, heat, electricity, etc. Such terms as imponderable matter, latent heat, electric fluid, forces of nature, and others in common use in text-books and elsewhere, served to maintain the distinctions; and even to-day some of these obsolete physical agencies are to be met in books and places where one would hope not to find them. As all physical phenomena are reducible to the principles of mechanics, atoms and molecules are subject to them as much as masses of visible magnitude; and it has become apparent that however different one phenomenon is from another, the factors of both are the same,—matter, ether, and motion; so that all the so-called forces of nature, considered as objective things controlling phenomena, are seen to have no existence; that all phenomena are reducible to nothing more mysterious than a push or a pull.

Some say that science is simply classified knowledge. To the author it is more than that, it is a consistent body of knowledge; and a true explanation of any phenomenon cannot be inconsistent with the best established body of knowledge we have. If physical factors are fundamental, then theorizers must square their theories to them.

The text-books have not kept pace with the advance of knowledge; and there is a large body of persons desirous of knowing more of natural philosophy, and especially of its trend, who have neither time nor opportunity to read and digest monographs on a thousand topics. To meet the wants of such, this book has been written. It undertakes to present in a systematic way the mechanical principles that underlie the phenomena in each of the different departments of the science, in a readable form, and in an untechnical manner. The aim has been to simplify and reduce to mechanical conceptions wherever it was possible to do so.

One may often hear the question asked, What is electricity? but a similar question as to the nature of heat or light or chemism is just as pertinent, although there chances now to be less popular interest in these than in the former; not, however, because they are in themselves better understood, or less interesting.

It is hoped that some of those whose interests lie along such special lines as chemistry, electricity, and even biology, will find something helpful in the chapters dealing with those subjects.

In covering so much ground in so small a treatise, it was necessary to select such facts as give prominence to fundamental principles. Doubtless others might have selected different materials, even with the same end in view, for otherwise competent persons are generally more familiar with certain details of a given science than with others; and I have used what was closest at hand.

Aside from the topics usually treated upon in a book of physics, the reader will find a chapter on Physical Fields, which is unique, as it extends the principle of sympathetic action—recognized in acoustics—to the whole range of phenomena, including living things.

The chapter on Life, in a treatise on physics, must justify itself; while the one on Machines points out their functions in a more complete way than has been done before.

Lastly, however large the physical universe may be, and however exact such relations as we have established may be, it is daily becoming more certain that even in the physical universe we have to do with a factor,—the ether,—the properties of which we vainly strive to interpret in terms of matter, the undiscovered properties of which ought to warn every one against the danger of strongly asserting what is possible and what impossible in the nature of things. With the electro-magnetic theory of light now just established, and the vortex ring theory of matter still *sub judice*, but with daily increasing evidence in its favor, one may now be sure that matter itself is more wonderful than any philosopher ever thought. Its possibilities may have been vastly underrated.

In the book called "The Unseen Universe," it is pointed out that possibly the ether may be the medium through which mind and matter re-act. What fifteen years ago was deemed *possible*, is to-day deemed *probable*, and to-morrow may be demonstrated; and a perusal of that book is recommended to persons who are interested in questions of that kind.

CONTENTS

MATTER, ETHER, AND MOTION

CHAPTER I

Matter and Its Properties

ALL kinds of phenomena that we can become conscious of through any of our senses are traceable directly or indirectly to what we call matter. The sense of feeling implies contact with a body of some kind; the sense of hearing depends upon movements of the air, which is a body of matter having certain properties; and the sense of sight, also due to vibratory motion, implies that matter exists, however distant, which has given rise to the vibratory motions that are perceived as light. So of taste and smell, actual contact of material particles endowed with particular properties are the conditions for exciting these sense perceptions. Some philosophers have added a sixth sense to the five senses we have recognized for so long a time—the sense of weight, as distinguished from the sense of touch; and still others have thought to distinguish a sense of temperature—relative perceptions of heat and cold, from the sense of touch; and if these truly represent distinct senses, they illustrate still further the truth that it is through the reactions of matter upon the nervous organizations of living things that all of our knowledge of things about us and of the universe as a whole is obtained.

It might seem to one as if our knowledge of matter should be tolerably good, accurate, and complete, seeing that it is thrust upon us everywhere, and affects us for good or evil continuously from the dawn of sensation till death; yet it may truly be said that the knowledge of matter, its properties, and the wonderful complexity of phenomena that are due to them, which we possess to-day was wholly unknown to all mankind until the time of Sir Isaac Newton, whose discovery of the law of gravitation was the first discovery of a universal property of matter; and by far the larger part of the knowledge we have, has been acquired in this century and mostly within the last half of it. The mass of mankind is, as it always has been, without any knowledge at all and without any desire for it. Whatever we have is due to the work of a small number of persons in Western Europe and America. Probably the large majority of mankind are quite unable to understand phenomena and their significance, yet among the brighter and more competent individuals in every country there is an apathy and indifference to the subject, due, of course, to the estimate they have of its degree of importance; and this estimate is in a good measure due to the philosophy of things in general held by the individual thinkers.

When Mr. Emerson was told by a Millenarian that the world was coming to an end the next day, he declared that he could get along without it, and so it probably has seemed to the majority of philosophers that the material world was a condition of things to be endured, rather than to be understood and utilized: that they were in it but were not a part of it.

Knowledge has, however, increased, and the wise ones are growing wiser; and some of the modern questions of philosophy and psychology are now so woven in with physical details that a knowledge of matter and its possibilities has become to them imperative.

There have been many attempts to define matter, such as, whatever occupies space, or whatever affects our senses, and so on; and there is no brief definition that has been generally adopted. In the ordinary affairs of life one rarely needs to make such distinctions as are necessary in philosophical and scientific affairs, where accuracy and clearness are of the utmost importance. There seems to be no way to define matter except by means of some of its properties. If we say that it is whatever occupies space, there is implied in the statement that the term is properly applicable to everything that exists in space; but so far as we know there may be any number of things in illimitable space that are not subject to any of the physical laws, such as a piece of wood or an air particle are known to be controlled by. If we say whatever affects our senses, we again are going beyond our warrant; for electricity is capable of affecting several of our senses,—sight, taste, feeling,—and yet there is no good reason for thinking electricity to be matter.

There is one property of matter that may seem to differentiate it from everything else, and hence, if adopted, will enable one to be precise about his use of the term. One part of the law of universal gravitation is—*every particle of matter in the universe attracts every other particle.* This makes gravitation a universal property of matter. The astronomers have observed the movements of exceedingly distant stars to be in accordance with this law, and there are no exceptions to it that have been discovered.

If, then, one adopts as the definition of matter, *whatever possesses the property of gravitative attraction,* he will have a definition that is in accordance with everything we know, and with the added advantage that if there be anything else in the universe that involves observable phenomena he will not need to confuse it with the phenomena of gravitative matter. This is the sense in which that term is used throughout this book.

Matter presents itself to our senses in a scale of magnitude from particles in the neighborhood of the hundred-thousandth part of an inch in diameter, and requiring the highest powers of the microscope to see, to such huge masses as that of the earth, eight thousand miles in diameter, the planet Jupiter, nearly eighty thousand miles, and the sun, eight hundred thousand miles in diameter, while some of the more distant stars are probably ten times larger than the sun. The large masses, however, are but collections of smaller ones, each particle bringing its own properties of whatever kinds they may be; and it does not appear that new qualities are developed by simply changing the distance between bodies. So the properties of matter may be studied exhaustively without employing specimens inconveniently large.

The thin stratum of gold spread upon cheap jewelry has all the characteristics and qualities of any specimen of gold however large; and a small test tube of hydrogen will exhibit all the kinds of phenomena that any larger quantity would show. For such reasons the study of the universe of matter can be carried on in the laboratory. The universe may be in the crucible one holds in the tongs; whatever difference there may seem to be, it will really be one of bigness only.

In treatises on physics one will generally find the properties of matter arranged in two divisions, called essential properties and non-essential ones. Of the former are (1) extension, or space occupying; (2) inertia, or passiveness under conditions of rest or motion; (3) impenetrability, or total and exclusive occupancy of its own space; (4) elasticity, or ability to recover its form after distortion, this, however, varying in degree in different bodies; (5) attraction, of which there are several varieties,—gravitation, acting at all distances; chemism, acting at close distances and selective in its operation, and apparently not existing at all between some kinds of matter, as, for instance, between oxygen and fluorine. Chemism is also capable of complete neutralization, and is thus in marked contrast with gravitative attraction, which is not affected in the slightest degree discoverable by contiguity; and lastly, cohesion, which is not apparent except bodies are in contact, but is the agency that holds the particles of bodies together so they form liquids and solids of any and all sorts.

The so-called non-essential properties are color, hardness, malleability, ductility, and the like, which vary very much in different substances. Among the metals silver is white, copper is red, gold is yellow. Diamond is the hardest substance known, while graphite is one of the softest, though both are composed of the same ultimate substance—carbon. Iron is malleable, and may be forged into any shape. Gold may be hammered out into leaves no more than one three-hundred-thousandth of an inch thick, but zinc is wholly unmanageable in that way. Platinum, one of the heaviest metals we have, can be drawn out into a wire finer than a spider's web,—a single grain may be drawn into a mile of wire; while bismuth, also a metal, cannot be drawn at all.

There are other conditions of matter that offer opportunities for convenient grouping sometimes, such as the solid, the liquid, and the gaseous: the solid being the one where the parts strongly cohere; the liquid, where the parts have but slight cohesion; and the gaseous, where the individual particles do not cohere at all, but, being elastic, bump against each other and rebound continually.

Farther on it will be shown how all substances may assume either of these conditions, inasmuch as it is temperature that determines whether a given substance be a solid, a liquid, or a gas.

Density signifies compactness of matter, or the relative number of particles in a given unit volume. If compression be applied to two cubic feet of common air until it occupies but one cubic foot, there is twice as much matter in that cubic foot as there was at the outset, and we express that fact by saying that the density is doubled. If twice the amount of matter is in the unit space, evidently the weight of the matter in that space must be twice what it was at first. So one may measure the density of matter by the weight of a unit volume of it compared with the weight of the same volume of some other substance taken as unity. Thus, if a cubic foot of water weighs 62.5 pounds, and a cubic foot of rock weighs 155 pounds, the density of the rock is $2\frac{1}{2}$, which means that it is $2\frac{1}{2}$ times heavier than water, and that the amount of matter in the rock is $2\frac{1}{2}$ times greater than that of the water. Such determinations have been made of all the different materials that could be found, and extensive tables have thus been constructed; but it is seen that the appeal is to gravitation, and presumes that every particle obeys that law, and that degrees of compactness of matter do not affect the law. Such comparative tables, based upon gravitation measure, are frequently called tables of *Specific Gravity*, so that density and specific gravity mean substantially the same thing. The following examples of the relative densities of bodies may be of interest:—

Gold,	19	Diamond,	4	Alcohol,	.8
Silver,	10.5	Common Stone,	2.5	Ether,	1.1
Copper,	8.8	Wood,	.8	Water,	1
Iron,	7.8	Sulphuric Acid,	1.8	The Earth,	5.6

Such numbers are to be understood as signifying that if a given volume of water weighs one pound, an equal volume of gold weighs nineteen pounds, an equal volume of iron seven and eight-tenths pounds, and so on.

Sometimes, however, it is convenient to choose for a standard of density some body, a small unit volume of which is much lighter than water, such as air, or more frequently hydrogen gas, a hundred cubic inches of which weigh 2.2 grains. In the metric system, a litre, which is nearly two pints is the standard of volume; and a litre of hydrogen weighs .0896 of a gram.

In chemical work this is the common standard for gases; while for solids and liquids a cubic centimetre of water is taken, which weighs one gram.

DIVISIBILITY OF MATTER.

Particles of matter as small as the hundred-thousandth of an inch may be seen with a good microscope as the smallest visible thing, but there is no reason for thinking that such a degree of fineness is any approach to the ultimate fineness of the parts into which it is possible to divide matter. For a long time philosophers have considered whether or not there could, in the nature of things, be an actual limit to the divisibility of matter, so that the smallest fragment could not be again divided into two or more parts by the application of appropriate means, thus making matter infinitely divisible, at any rate ideally.

In Mr. Spencer's "First Principles" this subject is considered at length, and the conclusion reached that it is impossible to conceive the existence of real atoms—bodies that cannot be divided into halves; nevertheless, we shall see presently that it is possible to conceive precisely that thing. It will be best here to note how far division has been carried and the means employed to effect it.

If a bit of phosphorus be put into a solution of gold, the gold will be set free in such a finely divided state that the particles remain suspended in the solution, giving to it a blue, green, or ruby color, depending upon the degree of fineness into which it has been broken up. Faraday estimated that the particles of gold in the ruby-colored liquid did not exceed the five-hundred thousandth part of the volume of the liquid. One-eighth of a grain of indigo dissolved in sulphuric acid will give a distinctly blue color to two and a half gallons of water, which would be about the millionth part of a grain to a drop of the water.

A grain of musk will keep a room scented for many years. During the whole of the time it must be slowly evaporating, giving out its particles to the currents of air to be wafted presently out of doors; yet in all this time the musk seems to lose but little in weight.

The acute sense of smell of the dog is well known; for he can detect the track of his master long after the tracks have been made, which shows that some slight characteristic matter is left at each footfall.

A spider's web is sometimes so delicate that an ounce of it would reach three thousand miles, or from New York to London. No one would think it likely that such a web would be made up of a single row of atoms, like a string of beads; for it would not seem probable that such a string could have such a degree of cohesion as spiders' webs are known to possess.

Chemists have concluded from their experience with matter in its various forms and conditions that it is really reducible to ultimate particles which have never broken up, no matter what conditions they have been subject to; and these ultimate particles are called *atoms*. The term is not now understood to signify what is implied in its derivation, as something that cannot be divided, only something that has not yet been broken up into smaller parts. Thus hydrogen, oxygen, iron, silver, are reducible to such ultimate atoms; and there are now known about seventy different kinds of atoms, and these are often spoken of as the elements. Though they are excessively minute when compared with ordinary objects of sight, yet they have a real magnitude which the physicist has measured in several different ways. Most of these methods are complicated, and, in order to be understood, require a pretty thorough knowledge of molecular physics; but the following one may probably serve to give one an idea of the degree of smallness which atoms must have.

When a soap-bubble is blown, the material of the film slides down the sides, making the bubble thinnest on top. When a certain degree of thinness has been reached at the top, colors begin to appear in concentric rings, and these colors appear to move towards the equatorial regions, new rings being formed at the top as fast as room is made for them by the displacement of the earlier ones. These colors always appear in the same order as they are in the rainbow, namely, beginning with the red and ending with the violet, then another set with the same order, until there have been ten or more such sets of rainbow tints. They are explained as being due to what is called interference in the light waves that fall upon the film. Light is reflected more or less from every surface it reaches. Some light is reflected from the first or outer surface of the film; some goes through the film to the inner surface, and is there reflected back to the outer surface, and then takes the direction that the light has which is reflected from the first surface, so that the light that reaches the eye from a point on a bubble comes from both outer and inner surfaces. That coming from the inner surface has had to travel farther than that coming from the outer surface by a distance of twice the thickness of the film. As light consists of waves, if one set of waves all of a length be made to move in the same direction as another set having the same length, their crests may coincide and produce a single higher wave; or the crest of one may be behind the crest of the other at any distance up to one-half the length of the wave itself, in which case the crest of one will coincide with the trough of the other, and the two waves will cancel each other, and this process is called interference. Now, in the case of the bubble, when the thickness is such that the distance through the film and back again is such as to equal half a wave length of a given kind of light, that particular wave is extinguished; and when one of the constituents of white light is wanting, that which is left is seen as colored light, and the color seen must depend upon the kind of color that has been cancelled. Red light has the longest wave length, about one forty-thousandth of an inch, and violet, the shortest of the waves we see, about one sixty-thousandth of an inch; and when these colors are seen upon the bubble we are assured that the interferences are produced by thicknesses due to fractional parts of such wave lengths. As the ray must go through the thickness twice in order to fall behind one-half of a wave, it follows that the thickness of the film where the last set of colors appear can be no more than one-fourth of the wave length of the shortest wave we can see, that is,

$$\frac{1}{4} \times \frac{1}{60,000} = \frac{1}{240,000} \text{ of an inch.}$$

When a bubble has reached this degree of thinness, so that no more colors are to be seen, a rather remarkable physical effect may be noticed. The film becomes almost jet black, with a jagged edge well defined between it and the brighter colored rings where the adjacent tint is purplish. The thickness of the film has fallen suddenly off here to about one-fortieth of the thickness it has where the tint is visible, and the bubble breaks in a second or two after this black patch appears; that is, when its thinness at any point becomes as small as

$$\frac{1}{240,000} \times \frac{1}{40} = \frac{1}{9,600000} \text{ of an inch.}$$

As the bubble, however, does persist for a short time, and the thin film has cohesion enough to enable it to support the weight of the bubble, it seems highly probable, but is not absolutely certain, that it must be more than one molecule of water thick at the thinnest place, which is, as shown, only about the one ten-millionth of an inch thick. If one thinks it probable that it be, say five molecules thick in order to have the degree of cohesion it shows, then the size of such molecule of water out of which the bubble is made can be but the one-fifth of the above small fraction, which gives about the one fifty-millionth part of an inch as the diameter of a molecule of water.

But a molecule is not the same thing as an atom: it is made up of atoms, chemically combined, and is defined generally as being the smallest fragment of a compound body that can exist and possess the physical characteristics that belong to such body. Thus, a drop of water possesses all the characteristics of any larger quantity of it, and a drop may be divided into smaller and smaller globules, perhaps a million of them, each one being visible with a good microscope; but if the division be carried to a higher degree, as it can be by various methods, chemical, electrical, and thermal, the qualities of water disappear, and two different substances, oxygen and hydrogen, are left, both gaseous under all ordinary conditions, and neither of them exhibiting any properties like water or from which any of the properties of water might be inferred. It may be well to remark here that this is only one illustration out of multitudes that might be named throughout the whole domain of physical science, that the properties of things under common observation are not simply the properties that belong to the elements out of which the things are built up; such properties being the result of collocation rather than inherent qualities.

The molecule of water is then a compound thing, and is made up of three atoms,—two of hydrogen and one of oxygen,—and therefore the actual size of an atom of hydrogen must be less than that represented by the above small fraction of an inch. Evidently a thing made up of three individual parts and two dissimilar substances cannot be spherical, and it will be well to bear this in mind in thinking of molecular forms. One may imagine the atoms themselves to be spheres, or cubes, or tetrahedra, or rings, or disks, or any other forms he likes, for the purpose of getting some sort of a mental picture of what a molecule might look like if it could be seen with a microscope; and it is probable that very many persons have hoped or thought that the microscope would sometime be so far perfected as to enable one to actually look upon the molecules of matter and perhaps upon their individual atoms. Let us therefore consider the problem of how much more powerful a microscope must need to be than any we possess to-day, in order that one should see a molecule! We will assume atoms to be about the one fifty-millionth of an inch in diameter, and that when combined into molecules they are geometrically arranged so that the diameter of a molecule made up of a large number of atoms is proportional to the cube root of the number of atoms, as is the case with larger bodies, say a box of bullets.

A molecule of water contains three atoms, a molecule of alum about one hundred, while, according to Mulder, a molecule of albumen contains nearly a thousand atoms. Then, according to the assumption, the molecule of alum would have a diameter equal to

$$\frac{\sqrt[3]{100}}{50,000000} = \frac{1}{10,776000} \text{ of an inch,}$$

and that of albumen would be equal to

$$\frac{\sqrt[3]{1,000}}{50,000000} = \frac{1}{5,000000} \text{ of an inch.}$$

Now, the best microscopes made to-day will enable one to see as barely visible a point the one hundred-thousandth of an inch, so that such a microscope would need to be as much more powerful than it now is as one hundred thousand is contained in five millions, that is, fifty times, in order to see the albumen molecule, and for the alum molecule as many times as one hundred thousand is contained in ten million seven hundred thousand, that is, one hundred and seven times. Now, one who is familiar with the microscope would probably admit that one might be made through improved methods of making and working glass hereafter to be discovered, two or three, or even ten times better than the best we have now; but the idea of one being made fifty or one hundred times more powerful than we have to-day, I do not think would be allowed to have any degree of probability. The case may be illustrated as follows: Suppose in the days of the stage-coach some one had imagined that by some improvement in methods of travelling one might some day travel one hundred times faster than the stage-coach could then go. Twelve miles an hour was not an uncommon rate then; but one hundred times that would be twelve hundred miles an hour, and that is sixteen times faster than the best we can now do, and about twenty-five times faster than express-trains now go. As a matter of fact, we travel about three or four times faster than the best stage-coaches did, and, on a spurt, may go six or eight times faster. The powers of the microscope have not been doubled within the last fifty years, and I suppose more time and ingenuity have been given to the problem of improving it than will ever be given to it in the same interval again.

There is another and still more serious reason why there is no probability that any one will ever see a molecule, even though the microscope had the magnifying power sufficient to reveal it; namely, the motions that molecules are known to have would absolutely prevent one from being seen. A free molecule of hydrogen has a velocity of motion at ordinary temperatures of upwards of a mile in a second, and its direction of motion is changed millions of times in a second. A microscope magnifies the movements of an object as much as it does the object itself. An object in the field of a microscope that should have a movement no greater than the hundredth of an inch in a second could only be glimpsed, so there is no possibility of one's being able ever to see a free gaseous molecule. Supposing one should be seized and held in the field, even then it is to be remembered that it is in a state of vibration, changing its form constantly on account of its temperature, so that its wriggling would prevent any inspection.

Lastly, there is every reason to believe that the molecules of all bodies are so perfectly transparent that they can no more be seen than can the air, even if there were no difficulty from their smallness and their motions.

If the atoms of a single element like hydrogen are so minute, so restless, and so transparent that no one can hope to see them so as to make out their forms and what gives them their characteristic properties, what shall be said of the case of seventy or more elements similarly minute and restless and transparent, yet each one easily identified in several ways, physical and chemical? Does it seem in any way probable that such differences in properties as are exhibited by gold, carbon, iron, and oxygen can be due simply to differences in size or shape of the atom? Presumably not; and the constitution of matter has therefore always been a mystery to philosophers, for if one is to attempt to philosophize upon the subject in accordance with such other knowledge as we have, one would need to conclude that if the different kinds of matter, the elements as we know them, were formed out of some prior kind of substance, as bullets and marbles are formed out of lead and clay, then there must be as many different kinds of substances out of which the different elementary atoms are formed as there are different elements, which proposition does not seem to have such a degree of probability that any one could adopt it. If one sought for the explanation of the different properties by assuming that all the different kinds of elements were formed out of one and the same fundamental substance, then it is equally difficult to understand how mere differences in size and shape could give such profound differences in quality as the elements possess.

Then, again, it appears that the individual atoms of each element are precisely alike. One atom of hydrogen is precisely like every other atom, so far as we have definite knowledge. Sir John Herschel likened them to manufactured articles on account of their exact similarity. A machine may turn out buttons or hooks or wheels or coins so exactly like one another that no one can tell them apart. It is really appalling to think of the immense numbers of atoms of every one of these seventy elements. It is a simple matter to calculate how many atoms there must be in say a cubic inch. It requires no other process than the application of the multiplication table. If the diameter of one be the fifty-millionth of an inch, then fifty million in a row would reach an inch, and a cubic inch would contain the number represented by the cube of fifty millions, which is

$$125000,000000,000000,000000,$$

(125 followed by twenty-one ciphers) a number which is more conveniently represented by 125×10^{21}. The utter impossibility of conceiving such a number will be apparent if one would try to represent to himself what the magnitude of only one million really is. Go out on a clear but moonless night and the heavens appear to be filled with stars. Count all that can be seen in a certain portion of the sky, say one-tenth, as nearly as can be estimated, and then determine the number in the sky that are in sight by multiplication. It will be discovered that only about two thousand can be seen in the whole sky. If one million stars were to be thus visible, it would require five hundred firmaments as large and as well filled as the one looked at to contain them. With the largest telescopes less than a hundred millions of stars are visible; but what shall one say when he learns that beyond a peradventure the number of atoms in a single cubic inch of matter of any sort is more than a million of millions times

all the stars in all the heavens visible in the largest telescope.

If one fancies that kind of work he may compute the number of atoms that make up the world. Of course it will make the number much larger; but when written out not so much longer as one might think, for when it is multiplied a million times it will add but six ciphers to it. Some mathematicians have been to the pains to compute the number of atoms there are in the visible universe, or, rather, the number that cannot be exceeded; for if the number stated above fills a cubic inch, if one knows the diameter of the visible universe, the space it occupies can readily be known in cubic miles and cubic inches, and if all this space was filled with atoms one could know and write down their number. Astronomers tell us that some stars are so distant that their light requires as long as five thousand years to reach us, although the velocity of light is as great as 186,000 miles in a second, and this distance is to be measured in every direction about us. If this be our visible universe, then the maximum number of atoms in it are calculable, and are stated to be represented by the figure 6 followed by ninety-one ciphers, or, as it is usually written,

$$6 \times 10^{91}.$$

If we return to microscopic dimensions, and compute the number of atoms, there will be in the smallest amount of matter that can be seen with the highest powers of the microscope, the one hundred-thousandth of an inch, it will be seen that five hundred atoms in a row would just reach the distance; and the cube of 500 is 125,000,000, that could be contained in a space so small as to appear like a vanishing-point and the structure or details be utterly invisible. We have read of spirits that could dance upon the point of a needle, but the point of a needle would be a huge platform when compared with this last visible point with the microscope; and the spirit that should dance upon it might be a million times bigger than an atom of matter, and not be in danger from vertigo. One may be astonished at the amount of intelligence associated with the minute brain structure of some of the smaller forms of animal life—say the ants; but from the above it will be seen that so far as such intelligence is associated with atomic and molecular brain structure, the size of the brain in the smallest ant, though measured in thousandth of an inch, is sufficiently large to involve billions of atoms, and the permutations possible are almost unlimited. The same idea is applicable to the brain of man, and seems to indicate that such differences in quality of mind as we see are not so much due to the differences in amount of brain, measured in cubic inches, as in atomic and molecular structure.

The work of physicists and chemists, carried on for many years, has convinced them that none of the processes to which matter has been subjected has affected its quantity in the slightest degree. A definite quantity of hydrogen, or, what is precisely the same thing, a definite number of hydrogen atoms, may be subject to any conditions of temperature, may be made to combine with other elements successively, forming with them solids or liquids or gases, and no atom is destroyed nor its individual properties changed in any degree. Neither has any phenomenon been discovered indicating that new atoms of any kind are ever produced by any physical or chemical changes yet known. Time does not alter them. Elements that have been imbedded in rocks from primeval times, reckoned by millions of years, when liberated to-day and tested, exhibit precisely the same characteristics as those obtained from other sources and that have been subject to many artificial conditions. Sometimes a meteorite reaches the earth, a sample specimen from distant space, having moved in some orbit about the sun for millions of years. Thousands of such bodies are in our possession, and they have been carefully analyzed, but no element unfamiliar to the chemist has been found among them; and the iron, the nickel, the carbon, the hydrogen, and all the rest of the elements that compose them, behave in every particular like those found on the earth.

So far as spectroscopic evidence goes, it testifies to the presence of the same elements in the sun and planets and comets; and it is as certain as anything physical can be, that the expert chemist here would be an equally expert chemist in the planet Mars, if he could find a way to cross the immense space that separates that star from us.

These facts and conclusions are frequently stated in such a form as this, namely, that matter cannot be created or annihilated. All that can fairly be meant by such language is that under all the conditions at present known, the quantity of matter remains constant; and this proposition has a high degree of importance in social affairs as well as in philosophy. If matter were liable to change in its quantity or quality by being subject to various physical conditions, all industries involving commercial interests would be in an unstable state. If the ton of iron ore should turn out, when smelted, only fifty per cent of iron instead of sixty per cent, as now,—the rest being either annihilated or transformed into lead or gold, or something else,—the smelting company would soon go bankrupt, even if gold were the product instead of iron, for if gold were liable to be produced in that kind of a way, its value would be next to nothing as a standard of value.

The old alchemists sought to transmute what they called the baser elements into gold. It is safe to say, if it were physically possible to do it and some one should discover the art, and it were an economical process, commercial disaster such as the world has never known would follow its announcement. It would be as if the volcanoes of the world should suddenly begin to eject gold in the place of lava.

Stability of physical properties is as essential for the stability of society as the regular recurrence of day and night; and philosophy would be impossible if fundamental data were not in every way immutable.

These physical principles lead to some curious and most interesting conclusions with regard to the great difference there is between bodies of matter of any and all kinds that are familiar to our senses and the atoms out of which these larger bodies are composed. In every case, where there is a difference in movement between two of these larger bodies made up of atoms, there is what we call friction, which invariably results in wearing away some of the material of both. It is the result of mechanical friction, to tear away some of the surface molecules of the two bodies. Bodies in use much, and therefore most subject to friction, become worn out. Our clothing is a familiar example; the journals of machinery, the tires of wheels, the sharpening of tools, the polishing of gems, the weathering of wood and stone,—all show that attrition removes some of the surface materials of such bodies, but there is nothing to indicate that attrition among atoms or molecules ever removes any of their material. It appears as if one might affirm in the strongest way that the atoms of matter never wear out, are not subject to such friction and the consequent destruction as comes to all bodies made up of them. The molecules of oxygen and nitrogen that constitute the air about us have been bumping and brushing against each other millions of times a second for millions of years probably, and would have been worn out or reduced, as the rocks upon the seashore have been beaten and ground into sand, if they had been subject to friction. So one may be led to the conclusion that whatever else may decay atoms do not, but remain as types of permanency through all imaginable changes—permanent bodies in form and in all physical qualities, and permanent in time, capable, apparently, of enduring through infinite time. Presenting no evidences of growth or decay, they are in strong contrast with such bodies of visible magnitude as our senses directly perceive. Valleys are lifted up and become mountain-tops; mountains wear away and are washed into the ocean; the beds of the ocean sink and rise; and the boundaries of continents may be worn and washed away through the incessant beatings of waves against their coasts. Wear and tear go on in all inanimate nature unceasingly, so that it is only a question of time when everything we see upon the earth will have changed beyond identification. The sun is shrinking, and must some time cease to shine. The stars, too, are changing likewise, because they shine, and their places in the firmament will be vacant. All living things grow because of change, and decay because of more rapid change, and there appears to be nothing stable but atoms. If it could be shown that life itself and the mind of man were in some way associated with atoms of some sort, as inherent properties, the hopes and longings cherished by mankind for continuous existence beyond the short term of three score years and ten would give way to convictions as strong as one has in any physical phenomenon whatever; the evidence would be demonstrative in the same sense as it is for the existence of atoms and their physical qualities.

CHAPTER II

The Ether

AN incandescent electric lamp consists of a fine thread of carbon fixed in a glass bulb from which the air has been exhausted. When a proper current of electricity is permitted to traverse the carbon filament, it becomes white-hot and gives out light like any other hot body. Other luminous bodies are in the air, and one might infer that the light was transmitted from the heated body to the eye by the material of the air itself. The light in the vacuum shows that this is not necessarily so, for the more perfect the vacuum is made the more freely does the light from the filament reach the glass bulb that encloses it. One is therefore led to infer that matter is not the agent that transmits light. The light of the sun reaches us after travelling through ninety-three millions of miles of space in about eight minutes. There are the best of reasons for believing that the atmosphere of the earth does not reach at most more than two hundred miles upwards from the surface, and its density at the height of only one hundred miles is such that there would be only about one molecule to the cubic foot.

It is not unlikely that there are free-roving molecules in space, as there are meteors in all directions about us, varying in size from fractions of a grain to masses weighing some tons, but the distance apart of these bodies is so great on the average that they cannot be considered as either help or hindrance to the passage of the light of either sun or stars. It is known with certainty that what we call the light from shining bodies is a kind of wave motion. The phenomena of interference, which can be brought about in several different ways, and which was referred to in the first chapter when speaking of the colors of soap-bubbles, show this. It is possible to annihilate two rays of light by making one of them to follow the other in a certain way; and one cannot conceive that two particles of matter of any sort could annihilate each other by simply changing their positions, but this is precisely what happens in light.

Wave motions of all kinds can cancel similar wave motions; for the wave consists of periodic movements, a crest and a trough, and when the crest and trough of one wave are superposed upon the trough and crest of another similar one, the result is the destruction of both waves. The lengths of these waves have been measured by a great many persons in various parts of the world, and they all concur that light can only be explained by wave motions such as they measure.

If there be wave motions, evidently there must be something moved. One cannot conceive of a wave movement when there is nothing that can be moved; so men have been compelled to believe that there is some medium between the sun and the earth that is capable of wave motion, and this medium they have agreed to call *the ether*.

If one admits the existence of ether between the sun and the earth as the agency for the transmission of light, he will need to do much more than that. The sun is but about ninety-three millions of miles distant, but most of the planets are hundreds of millions and some of them thousands of millions of miles from us, and the light comes from them too; so the ether must extend through the space occupied by the solar system, the diameter of which is six thousand millions of miles, and to cross this space light requires nine hours, though going at the rate of one hundred and eighty-six thousand miles per second.

Then there are the stars beyond our solar system, the nearest one so distant as to require three and a half years for the light to get to us at the same rate; and some of these are so remote that thousands of years are needed for their light to arrive. That light we see from them to-day left them before America was discovered, before Jesus was born, before the pyramids were built, and for all we should be able to see they might have ceased to exist long ago, though their light continues to shine. So the ether must extend to those most distant stars we can see, and that, too, in every direction. There is no exaggeration in the statement that our visible universe is so great that light requires ten thousand years to cross its diameter. There is no reason, either, for setting that as a boundary to its magnitude; but wherever light comes from to us, there must this medium, the ether, be.

But there are other and just as good reasons for thinking there must be some medium between bodies, even when all atoms and molecules have been removed. For instance, everybody knows that one magnet affects another at a distance from it, and there is no kind of substance known that will prevent such action when interposed between them.

If one of these magnets be placed in the most perfect vacuum that can be made, it still acts as it would in the air, only with still greater freedom. One cannot believe that one body can thus act upon another body without some kind of a medium between them. Is it not absurd to think otherwise? One may, if there appears to him to be a good reason, suppose that there is a magnetic medium or ether different from that one employed in the transmission of light; but there is a similar need for imagining one for the effects produced by electrified bodies upon other bodies in their neighborhood. An electrified glass rod will attract a pith ball or anything else just as well in a vacuum as out of it; and it is certain that electrical attraction and magnetic attraction are not identical, for an electrified body will attract one kind of thing as well as another, while a magnet is selective in its effects, and affects iron chiefly. Hence, if each different effect in a vacuum is to be attributed to some different kind of medium, there would need to be an electric ether in addition to the other two.

Then there is gravitative attraction, which has before been mentioned. If it is not rational to think that one body can act upon another body not in contact with it and without some medium between them, then one is bound to admit that the gravitative effects observed, say between the moon and the earth, the sun and the earth, and in every other case, are due to the action of some medium between them. Neither is it at all needful to be able to explain *how* the medium acts thus and thus, or even to imagine how it might, in order to firmly believe that there must be one.

Here are four cases of apparent action at a distance

of one body upon another, requiring some sort of an intermediate agency; and, unless there be some good reason for thinking there are several such media occupying the same space apparently, it is much more philosophical to believe it likely that one medium exists capable of transmitting effects of the different kinds; and especially will this appear to be truer if it is known, as it is known, that the magnetic and electric effects are transmitted with the same velocity as is the light. So that physicists to-day quite concur in the belief that what was called at first the luminiferous ether, on account of its function in transmitting light, is the same medium that is concerned in the other phenomena of magnetism, electricity, and gravitation.

It is likewise true that there are some physicists who hold rather lightly upon this belief, taking it as a convenient working hypothesis, and who would seem to be ready in a minute to surrender the idea, unless it had been demonstrated in the same way as the existence of matter and of motion has been. But this is not the attitude of philosophic minds.

Sir Isaac Newton deduced from the observed motions of the heavenly bodies the fact that they attract each other according to the law now known as the law of gravitation, but he says nothing about *how* bodies can affect each other. That is, in his "Principia" he does not attempt to explain gravitation. He explicitly does say, however, that he has not employed hypotheses in his work, yet we know from other of his writings that the idea of a medium was constantly in his mind. His "Principia" closes thus:—

> "And now we might add something concerning a most subtle spirit which pervades and lies hid in all gross bodies; by the force and action of which spirit the particles of bodies mutually attract one another at near distances and cohere if contiguous; and electric bodies operate to greater distances as well repelling as attracting the neighboring corpuscles, and light is emitted, reflected, inflected, and heats bodies; and all sensation is excited, and the members of animal bodies move at the command of the will, namely, by the vibrations of this spirit mutually propagated along the solid filaments of the nerves from the outward organs of sense to the brain, and from the brain to the muscles. But these things cannot be explained in few words, nor are we furnished with that sufficiency of experiments which is required to an accurate determination and demonstration of the laws by which this electric and elastic spirit operates."

This shows plainly enough that he believed that some medium, different from matter, was essential for a mechanical conception of the phenomena he alluded to. In a letter to Bentley he states his philosophical judgment upon the subject in still stronger terms, and it shows, too, the sense in which he is to be understood when he says: "I frame no hypotheses"—which has frequently been repeated to adventurous hypothecators as the example of the model scientific man. Hear him!

"It is inconceivable that inanimate brute matter should, without the mediation of something else which is not material, operate upon and affect other matter without mutual contact, as it must do if gravitation in the sense of Epicurus be essential and inherent in it. ... That gravity should be innate, inherent, and essential to matter so that one body can act upon another at a distance through a vacuum, without the mediation of anything else, by and through which their action and force may be conveyed from one to another, is to me so great an absurdity that I believe no man who has in philosophical matters a competent faculty of thinking can ever fall into it."

Newton uses the word *Spirit* in the sense of a substance entirely different from matter (see page 36). Evidently Newton was so strong a believer in the medium that we call the ether, though he could not work out its mode of action, that he was ready to discount the intelligence of any man who doubted it.[1]

If our knowledge of the existence of the ether is

---

[1]In 1708 Newton wrote thus: "Perhaps the whole frame of nature may be nothing but various contextures of some certain etherial spirits or vapors, condensed, as it were, by precipitation; and after condensation wrought into various forms, at first by the immediate hand of the Creator, and ever after by the power of nature."

These with his other acute remarks concerning what we now call the ether lead us to infer that his mechanical instincts were more to be trusted in this field than his more labored efforts.

not so positive as it is for matter, but is inferential, it will be readily understood that the knowledge we have of its properties cannot be very exhaustive. Some have imagined that it was only a finer grained kind of matter than that we know as the elements, and that it must be made up of atoms, though almost infinitesimal in size. Others think it cannot be granular at all, but forms a continuous substance throughout space. By "continuous" is meant that there are no interstices in it: that it is constituted like a jelly, only not made up of distinct parts or atoms, so there can be no such thing as separating one part from another, leaving a vacuous cavity or rent between them. One of the reasons for thinking this to be the case is, that if it were made up of finer atoms or of atoms at all, such waves as those of light could not be transmitted by it. Longitudinal waves, like those of sound in air, can be transmitted by atomic or molecular structures but not transverse waves, that is, such as are at right angles to the direction of propagation. Some of these light waves are as short as the hundred-thousandth of an inch, and some are as long as the one two-thousandth of an inch, and perhaps longer. Yet all of them are transmitted with the same velocity in any and every direction. From the fact that light travels with the same velocity in every direction, it is inferred that the ether is not only homogeneous, but its properties are alike in every direction. As light is transmitted in straight lines, it seems to follow that there is no difference in its quality in different parts of space.

That wave motions travel with such high velocity in it has been interpreted as proving it to have a high degree of elasticity, while the fact that it offers no appreciable resistance to the movements of bodies of matter in it is supposed to indicate that its density is very small.

There are some, however, who think that such terms as elasticity and density are not appropriately applied to the ether. These terms signify properties of atoms and molecules. If density signifies compactness of atoms, then the word could not apply to something not composed of atoms. In like manner, if elasticity means ability to recover form after deformation, then it is not applicable to substances that cannot be deformed, and it is customary to speak of the ether as being incompressible. Still, it is certain that stresses may be set up in it in various ways, and that these conditions may be propagated, in certain cases in straight lines, in other cases in curved lines, so whether the explanation be forthcoming or not, there is no doubt about the facts.

There is no evidence at all that the ether is subject

to gravitative action, or that it offers any resistance to a body moving in it. That is to say, it gives no evidence of friction. Here is the earth rotating upon its axis, and the velocity of rotation at the equator is a thousand miles an hour, and if there were an appreciable amount of friction the earth must slowly be coming to rest like a top spun in the air. Yet the astronomers tell us that the length of the day has not changed so much as the hundredth of a second within the last two thousand years. Again the earth revolves in its orbit about the sun at the average rate of nineteen miles a second, and if the ether through which it moves offered any resistance to the motion, the length of the year would be changed, but no such change has happened in historic times. Again, such bodies as comets move very much faster than the earth; some have been known to have a velocity of three hundred miles per second when near the sun, but the comets complete their circuits and give no evidence of slackened speed due to friction in space.

If, then, the ether *fills* all space, is not atomic in structure, presents no friction to bodies moving through it, and is not subject to the law of gravitation, it does not seem proper to call it matter. One might speak of it as a substance if he wants another word than its specific name for it. As for myself, I make a sharp distinction between the ether and matter, and feel somewhat confused to hear one speak of the ether as matter.

Nearly thirty years ago Helmholtz investigated, in a mathematical way, the properties of vortical motions, and, among others, pointed out that if a vortical motion was set up in a frictionless medium, the motion would be permanent, and it could not be transformed. Sir William Thomson at once imagined that if such motions were set up in the ether, the persistence of their form and the possibility of a variety of motions would correspond very closely with the properties that the atoms of matter are known to possess. Such vortical motions as are here alluded to, all have seen, as they are often formed by locomotives when about starting, if the air be quiescent. Horizontal rings, three or four feet in diameter, may be seen to rise wriggling into the air sometimes to the height of several hundred feet. They may be formed also by smokers by a vigorous throat movement forcibly puffing the smoke from their mouths, and they can be made artificially by providing a box having a hole on one side an inch or two in diameter and the side opposite covered with a piece of cloth. A saucer containing strong ammonia water and another with strong hydrochloric acid may be set inside, and dense fumes will fill the box. If the cloth

DIAG. 1.

be struck by the hand, a ring will issue from the hole, and may go forward several feet, and its behavior may be studied. Such as are formed in the air under such conditions present so many interesting phenomena that it is worth the while here to allude to them for the sake of helping the mind to a clearer idea of how some of the properties exhibited by matter may be accounted for.[1]

1. The ring once formed consists of a definite amount of the gaseous material of the air in a state of rotation, and in its movements afterwards retains the same material. It is to be noted that the ring is formed in the air, the white fumes serving merely to make the ring visible. The ring moves forward in a straight line in the direction it is started, just as if it were a solid body. It may move very fast too,—ten feet a second or more, and reach the distant side of the room, but it always moves of its own motion in a direction perpendicular to the plane of the ring.

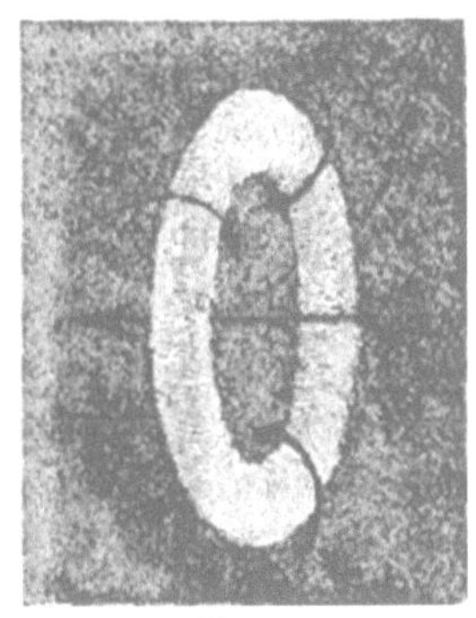

DIAG. 2.

[1]The method of producing these vortex rings and their phenomena are fully explained in "The Art of Projecting." By Prof. A. E. Dolbear. Illustrated. $2.00. Published by Lee and Shepard, Boston.

2. It possesses momentum, and will push against the object it hits.

3. If made to move rapidly adjacent to a surface like a wall or table, it will move towards it as if it were attracted by it, and generally will be broken up by impact against it.

4. A light body, like a feather or thread, will be apparently pushed out of the way in front of it, and drawn towards it if behind it—phenomena like attraction and repulsion.

5. If two such rings bump together at their edges, each one will vibrate with well-marked nodes and loops, showing that, as rings, they are elastic bodies, and that their period of vibration depends upon the rate of the rotation.

6. If two such rings be moving in the same line, but the hindmost one swifter so as to overtake the other, the foremost one enlarges its diameter while the hinder one contracts until it can go through the former, when each recovers its original dimensions.

7. If two meet in the same line, going in opposite directions, the smaller one goes through the larger and may be brought to a standstill in the air for a short time until the other has got some inches away, when it starts on in the same direction as before.

8. If two similar ones are formed at the same time,

side by side, at a distance of an inch or two, they always collide at once as if they had a mutual attraction. The result of the collision may be the destruction of one or both, or—

9. Each one may break at the point of impact, and the opposite ends may weld together, forming a single ring which will move on as if it had been singly formed, or—

10. Instead of breaking they may rebound from each other, but always at right angles to the plane in which they were moving at first; that is to say, if they were moving in a horizontal plane before impact, they will rebound from each other in a vertical plane.

11. Three rings may in like manner be made to join into one.

12. The material of the ring may often be seen to be in rotation about the ring, while the ring, as a whole, does not rotate at all, a rotary wave.

13. The parts of a ring may be in a state of vibration in the ring without changing its circular form, somewhat as if the ring were tubular and two bodies should move up on opposite sides till they met and rebounded to meet below, and so on.

All these, and some other just as curious phenomena, may be observed in vortex rings, and may fairly be said to be due to the properties of the rings themselves. For instance, the vibratory motions alluded to in the fifth show that elasticity is a property of the ring, and that the degree of elasticity does not depend upon what the ring is made of, but upon the kind and degree of motion that constitutes the ring. If such a ring could be produced in material not subject to friction, none of the motion could be dissipated, and we should have a permanent structure, possessing several properties such as definite dimensions, volume, elasticity, attraction, and so on, all due to the shape and motions involved.

Imagine, then, that vortex rings were in some way formed in the ether, constituted of ether. If the ether be, as it is generally believed to be, frictionless, then such a thing would persist indefinitely: it would have just that quality of durability that atoms seem to possess. It would possess physical attributes, form, magnitude, density, energy, that is, it would not be inert. It would be elastic, executing a definite number of vibrations per second. This property of elasticity has generally heretofore been assumed to be a peculiar endowment of ordinary matter, and one was at liberty to imagine some matter without it because not so made. This view implies that elasticity is a necessary property of vortex rings; for as the velocity of rotation is reduced, so is the degree of elasticity, and if there was simply a ring without being in rotation, it would have no elasticity at all, neither would it have any qualities different from the medium it was imbedded in.

That such a quality as elasticity may be due solely to motion, and varying with it, one may assure himself with that piece of apparatus to be found in most collections in schools known as Bonnenburger's. It consists of a disk of metal, mounted in gimbals so it can be set spinning in any plane. If this be set spinning in a vertical plane it becomes tolerably rigid in that plane, and cannot be moved out of it but by the employment of quite a degree of pressure. If the framework be quickly struck by the finger while thus spinning, the wheel will begin to rock back and forth like the prong of a tuning-fork, and the more rapid the rotation the higher the rate of vibration. When the velocity of rotation becomes slow the vibratory motion may be as slow as once a second, and, of course, when the ring is not revolving it will not vibrate at all. Thus there is fairly good physical reason for thinking that what we call elasticity in the atoms of matter may be due simply to the motion they possess,

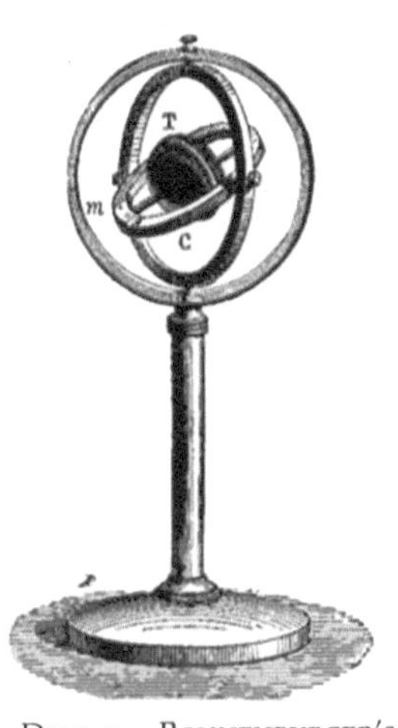

DIAG. 3.—BONNENBURGER'S APPARATUS.

and *how* that may be one can understand if atoms be vortex rings.

One may properly ask how one vortex ring can differ from another so there could be so many as seventy or more different kinds of atoms. To this it may be said that such rings may differ from each other not only in size but in their rate of rotation: the ring may be a thick one or a thin one, may rotate relatively fast or slow, may contain a greater or less amount of the ether. The word "mass" in physics is used to denote a quantity of matter as measured by its resistance to pressure tending to move it as a whole. Thus if a pressure of one pound be applied to two different bodies for say one second, and one of them was moved an inch and the other but half an inch when otherwise they were alike free to move, we would say that one had twice the mass of the other—its resistance to being moved was twice as great as the other.

In the case of the Bonnenburger's rotating disk, the resistance to the pressure tending to move it depends upon the rate of rotation, and a thin and swift moving disk would offer much greater resistance than a much larger one with a slower speed. So one might infer that the difference in what is called mass among the atoms of matter might be due simply to the different speeds with which the rings rotate, rather than in the absolute volume of ether in the state of rotation. There are other reasons than these for thinking that motion is the chief characteristic of matter. Chemists have discovered that both the chemical and physical properties of all kinds of matter are functions of their mass or relative atomic weights, and that they may be arranged in a harmonic series. Harmonic relations may imply either relations of position or of motion. But the fundamental properties of matter do not change by changing its position, and one is therefore led to the conclusion that one must look to the various kinds of motion involved among atoms for the explanation of all their properties and all their phenomena.

There is another very important and peculiar property possessed by vortex rings; viz., there cannot be such a thing as half a ring or any fragment of one. Break such a ring in two and there is not left the two halves; not only is the ring broken, but each part at once vanishes into the indistinguishable substance that composed it, and all the properties that characterized it as a ring have vanished with it.

This greatly aids one to understand that matter may not be infinitely divisible. Over and over again have philosophers asserted that it was impossible to imagine an atom of matter so small that it could not in imagination be again broken into two or more parts. A vortex ring, however, shows how the thing can be done. If an atom be a ring, when it is disrupted it is at once dissolved into ether, and that is the end of it. There are no fragments of the ring.

One, however, must not infer from the above treatment that it represents knowledge of a demonstrated kind, for it does not. It was remarked in the first chapter that atoms are too minute to be seen and studied as one would study an animalcule or a blood corpuscle, and one's knowledge must be altogether inferential concerning them; but what knowledge we do have, and the inferences that may properly be drawn from it, all tend to convince one that matter and the ether are most intimately related to each other, and that some such theory as the vortex ring theory of matter must be true.

Now, it is either that theory or nothing. There is no other one that has any degree of probability at all. If what is presented herewith is not the precise truth concerning a most difficult subject, it may have the merit of helping one to conceive the possibilities there may be of deducing qualities from motions, and rid him of the idea that matter consists necessarily of some created things that have no necessary relations to the rest of the universe beyond the properties impressed by fiat.

In the latter case one could never hope to understand them, because there could be no *necessary* reason for their being as they are, rather than some other way, whereas, in the former case, the mechanical relations can be understood, and there is left the possibility that by and by, with more light and knowledge, one may know the physical conditions under which matter itself came into existence.

CHAPTER III

Motion

EVERYBODY has so clear a conception of motion that there would not seem to be any difficulty in defining it absolutely, but philosophers and others from remote times till now have been perplexed by its problems. How can Achilles ever overtake the tortoise, though he runs ten times faster? How can the top of a cart-wheel move faster than the bottom? If the sun cannot set above the horizon and cannot set below it, how can he set at all? In the last chapter some phenomena were alluded to which were attributed to motions of different kinds, and one must needs have a definite notion of what he is talking about in order that his words shall convey to himself, as well as others, the information he would impart. Rest and motion are contrasted conditions of bodies, so if a body is at rest we say it is without motion, and *vice versa*. If two persons sit side by side in a house they may be said to be at rest, but if they sit side by side in a railroad car they will be at rest relative to each other as they were before, but may be in motion with reference to things outside the car. If, as a vessel sails past the end of a wharf, a person on board would talk with a person standing upon the wharf, he will walk so as to keep opposite the man standing still, and the two will be at rest in relation to each other, while one will be in motion with reference to everything on board the vessel. Thus it appears that rest and motion are relative terms, and can only be understood to apply to the relative continuous position of two bodies or objects. Hence, if there were but one object in the universe there could be no such thing as change of position, for that implies another body with which position may be compared at intervals. But such a single body might have some internal motions by which there was a relative change of position of its parts with reference to themselves. For instance, a tuning-fork might be at rest as a whole with reference to all other bodies, yet its prongs might vibrate towards and away from each other, the centre of mass or the centre of gravity of the fork itself not moving in the slightest degree either with reference to itself or anything outside itself. Again, a body might spin like a top, and there would be no change of position of the body as a whole with reference to any other body, nor change of position of the parts with reference to each other, yet there would be a change of position of the parts with reference to all bodies outside itself. Hence, a brief definition of motion is not so easy to give.

One might say that motion was the change of position of a body with reference to other bodies, or the change of position of the parts of a body with reference to each other, or the change of position of the parts of a body with reference to other bodies. But these would not cover all possible cases. There need be no trouble, however, in particular cases, because there will always be data at hand to determine the character and direction of the motion.

One may study the geometry of positions and changing positions of mathematical points, and attend only to rates and direction of motion of all sorts, without considering the motions of bodies of real magnitude possessing physical properties like matter. The science that has to do with such ideal conditions is called *kinematics*. Whenever the motions of matter are considered, the science is called *kinetics*. Of course all phenomena involve the motions of matter. Although one sees a great variety of motions, a few examples of particular sorts may be helpful in analyzing them.

1. The drifting of clouds, the flight of birds, of arrows, of bullets, of meteors, the sailing of vessels, the running of locomotives, are examples of one kind of motion; namely, where the change of position is that

of the body as a whole with reference to other bodies external to it. The cloud may drift with the air, but with reference to the surface of the earth it moves. Where a body thus moves straight on continuously with reference to other bodies, whether the distance moved be long or short, the motion is called *translatory* or *free-path motion*. The latter term is most frequently applied to the movements of the molecules of a gas. In ordinary air the distance apart of the molecules is on the average about the one two-hundred-and-fifty-thousandth of an inch, but the molecules themselves being only one fifty-millionth of an inch in diameter, it will be seen that they have a space to move in about two hundred times their own diameter before coming in collision with another one; and after collision their direction is only changed when they go on to another collision, and we say that their free path is on an average about the two-hundred-and-fifty-thousandth of an inch. With some modern air-pumps it is possible to reduce the amount of air in a space so that the average free path of a remaining molecule will be a foot or more; but neither the size of the moving body, nor the distance hundred times their own diameter before coming in collision with another one; and after collision their direction is only changed when they go on to another collision, and we say that their free path is on an average about the two-hundred-and-fifty-thousandth of an inch. With some modern air-pumps it is possible to reduce the amount of air in a space so that the average free path of a remaining molecule will be a foot or more; but neither the size of the moving body, nor the distance it moves, nor the velocity with which it moves, makes any essential difference in the specific kind of motion: so the movements of air particles among themselves, of billiard-balls between impacts, of a bullet on its way to the target, and of a planet or comet in its orbit, are all examples of the same kind of motion, namely, translational.

2. The swaying of the branches of trees when moved by the wind, the swinging of the pendulums of clocks, the movement of the piston in a steam-engine, of the prongs of tuning-forks, the reeds and strings in musical instruments, are examples of a different kind of motion, inasmuch as the changes of position relate to the body itself rather than to external bodies. The tuning-fork is the type of them all, and together they are called *vibratory* motions. Sometimes, when the bodies that move thus are large and the motion conspicuous, as, for example, in the pendulum of the clock, and the steam-engine piston, the motion is spoken of as *oscillatory*. In such cases, as in the former one, it should be borne in mind that mere differences

in the size of bodies, or of the rate of motion, does not in any manner change the character of the motion, so the name that is applicable to one will be equally applicable to all. If one calls the movement of a vibrating tuning-fork *vibratory,* the same term may be applied to an atom if it goes through a like periodic change of form, for that is the chief characteristic of vibratory motion; and hereafter it will appear how needful it is to bear this in mind, for what a given amount of motion will do will be seen to depend altogether upon the kind of motion.

3. The spinning-top, the balance-wheels of engines, the wheels of machines of all kinds, the turning of the earth, and each member of the solar system upon its axis, are examples of another sort, where the displacement is not, as in the last, between parts of the same body, but a change in the relative position of each part of a body to what is outside itself. The pendulum of a clock swings to and fro, but its point of suspension does not move; whereas every part of a turning-wheel is presented to opposite parts of space in the plane of its revolution. This motion is called *rotary,* and just as in the other two cases, I wish to emphasize the fact that the term is properly applicable to masses of matter of all degrees of magnitude; so an atom may spin on its axis as well as the earth

or sun, and the phenomena it will be competent to produce by such spinning will be very different from that produced by its vibrations or free-path motions.

These three kinds are all of the primary ones: all the others we see are made up of these or their compounds. For instance, a compound of a free-path motion with a vibratory motion will give a wave or sinuous motion if the direction of the vibration be at right angles to the free path. A combination of a free-path with a rotary may give a spiral motion, as illustrated by the movement of a screw when pushed and turned into a piece of wood.

In a sewing-machine may be seen all of these kinds of motion and some other compounds more complex than the ones spoken of, but one may readily analyze them into the three primary ones.

These forms of motion have been spoken of as if they were peculiar to matter; but it ought not to be inferred that motion is not attributable to the ether. Indeed, we know that some sorts of motions are propagated in the ether. For instance, what we call light is an example. Its form is *undulatory*; and, as we have seen above, an undulatory motion is a compound of a rectilinear and a vibratory. A spiral movement in the ether is also known, and it is sometimes called rotary-polarized light: its motion is like that of a screw, and we know that such a motion is a compound of a rectilinear and a rotary. Rotary motions in the ether are also known as taking place in front of magnetic poles, and are the results of the magnetism imparted to the iron or other substance. I am not aware that any simple rectilinear motion is known to occur in the ether: there may be, and likely enough is, such.

For convenience, motions that are large enough to be visible are called *mechanical motions*, while those too minute to be seen are often called *molecular* or *atomic*. Sometimes these molecular and atomic motions are spoken of as if they were mysterious, and not to be understood in the same sense as the larger ones that are visible to us; but it is difficult to justify any such distinction, and difficult to imagine that any kind of a motion a large piece of matter may have, a small particle or atom cannot have, and *vice versa*. It would seem probable that whoever finds a difficulty in this cannot have strong mechanical aptitudes, and is not gifted with an adequate scientific imagination.

A free body of any kind and of any magnitude may have any kind of a motion whatever, and may move in any direction and with different velocities, but the term velocity is used in different senses when applied to different kinds of motion. Thus the velocity of an atom in its free path, of a musket-bullet, of sound-waves, is measured in feet per second. The velocity of vibrating bodies is indicated by the number of vibrations they make per second. A tuning-fork making two hundred and fifty-six vibrations in a second is said to have that rate of vibration, whether the actual distance moved be one distance or another, which, of course, will depend upon the amplitude of each individual swing; while rotational velocity is generally specified by giving the number of rotations per second, or per minute, or some other unit interval of time. A top may spin a hundred times a second, a balance-wheel of a steam-engine turn four times, while the earth makes one revolution in a day of twenty-four hours. The range in velocities of these different kinds that have been measured is very great indeed. In free-path or translational motion, there may be the snail's pace, perhaps less than an inch a minute, the pace of a man walking say three miles an hour, which is at the rate of eighty-eight feet per minute. A race-horse may trot a mile in two minutes and ten seconds, which is forty feet per second. A steam-locomotive may run seventy miles an hour, which is nearly one hundred feet per second. A rifle-bullet may go a thousand feet, and a cannon-ball two thousand feet in a second. The earth in its orbital motion goes seventeen miles per second; meteors come to the earth, from space, sometimes having a velocity of fifty or more miles per second, while comets may reach the velocity of nearly four hundred miles in the same time when near the sun. These are the velocities of bodies of visible magnitude, but some of the motions of molecules are fairly comparable with some of these. Thus a molecule of common air is moving in its free path about sixteen hundred feet per second, while a molecule of hydrogen, which is much lighter, goes more than six-thousand feet—upwards of a mile—in the same time. As remarked before, the free path for air molecules having but about the two-hundred-thousandth part of an inch, it must change its direction an enormous number of times in a second,—as many times as one two-hundred-and-fifty-thousandth of an inch is contained in sixteen hundred feet;

$$250,000 \times 12 \times 1,600 = 4800,000000.$$

Four thousand eight hundred millions of times. How one may assure himself that such a statement is not fabulous will be pointed out farther on; so far one needs only to trust the multiplication table.

For vibratory rates there are also enormous ranges: there are the slow oscillatory movements of swinging pendulums of various lengths, sometimes occupying several seconds for the execution of one vibration; piano-strings having a range from about forty per second to four thousand; the chirrup of crickets about three thousand. Short whistles and steel rods have been made that will make as many as twenty thousand vibrations per second,—a rate much higher than can be perceived by most persons, though occasionally abnormal hearing in an individual enables him to hear sounds to which ordinary ears are entirely deaf. When the number of vibrations per second becomes so great that they cannot be individually seen nor heard, one must trust his judgment and the properties of matter in determining whether there really are any still more rapid. It has been found by experiment that the number of vibrations a given body can make when it is struck so as to produce a sound depends upon its shape, its size, its density, and its degree of elasticity. If a steel rod, having a given diameter and length, makes, when struck, five hundred vibrations per second, another similar one with but half the length will make twice as many in the same time. If one were made of something still more elastic than steel, and of the same size, the vibratory rate would be higher still.

A steel tuning-fork three inches long may make five hundred vibrations per second; if it were only the one fifty-millionth of an inch long it would make not less than 30000,000000 vibrations per second; and if it were made of a substance like ether it would make as many as 1000,000000,000000—a thousand million of millions per second. As large as this number is, and as improbable as it would seem to be, there is indubitable evidence that the atoms of matter do actually make such a number of vibrations per second.

If one knows the rate at which vibrations are propagated in a medium and the wave length, one can readily determine the number of vibrations the body is making that sets up the waves. Thus, if the velocity of sound in the air be 1100 feet per second, and the length of one wave be 1 foot, then the body must be making $\frac{1100}{1} = 1100$ vibrations per second: that is, the velocity divided by the wave length will give the number of vibrations.

The velocity of light is known to be 186000 miles per second; the wave lengths of light are also known with great precision, and are all only small fractions of an inch. If they were only one inch long, their number would be the number of inches there are in 186000 miles, or $12 \times 5,280 \times 186000 = 11784,960000$ per second. In reality they are only one forty-thousandth or the one fifty-thousandth of that.

$$11784,960000 \times 50000 = 589,248000,000000,$$

nearly six hundred millions of millions per second. No one can pretend to comprehend such a number; but in proportion as he understands the process and the data by which such a result is reached, will he have an abiding confidence that it is legitimate and that it expresses the actual truth.

Sometimes it is convenient to know the actual space that is moved over by a vibrating body in terms of free-path or translatory motion, that is, how far would the body move in the same time if, instead of vibrating, it went on in a straight line. If the prong of a tuning-fork moves through the one-hundredth of an inch each swing, and vibrates one hundred times in a second, obviously its rate of motion measured that way would be only one inch, which would be a relatively slow motion when compared with many others. If the same computation be applied to atoms, however, whose rate of vibration is so enormously high, it leads to some very respectable translational velocities. Thus, suppose the amplitude of vibration of an atom of hydrogen be as great as one-half its diameter, that is, one hundred-millionth of an inch, if it vibrates five hundred millions of millions of times per second, the actual space moved through will be

$$\frac{500,000000,000000}{100,000000} = 5,000000 \text{ inches} = 80 \text{ miles},$$

which is more than four times that of the earth in its orbit. It does not appear probable, however, that the amplitude of motion is anywhere near as much as that assumed, at any rate for ordinary temperatures; but if it be only the one-hundredth of that amplitude the velocity exceeds that which can artificially be given to any visible object, as it will then be nearly a mile a second.

Rotary speeds have wide ranges. The earth takes twenty-four hours to make one revolution; the moon about twenty-eight days, and the sun twenty-six, and some others of the planets perhaps much longer than that. Some astronomers have concluded from their observations of the planets Venus and Mercury, that their axial rotation corresponds with their time of revolution about the sun, being 224 days for Venus, and 88 for Mercury. Tops have been made to spin eight hundred or a thousand times per second; and if molecules ever rotate their rate has not been measured. The velocity of rotation, when measured as a translation, must evidently depend upon the diameter of the body rotating. The diameter of the earth being nearly eight thousand miles, a point on the equator moves twenty-five thousand miles in twenty-four hours—something over a thousand miles an hour, or about seventeen miles a minute. A driving-wheel of a locomotive that is six feet in diameter will advance nearly nineteen feet every revolution. To have a speed of a mile a minute, which is 88 feet per second, it must turn round $\frac{88}{19} = 4.6$ times per second. A disk 4 inches in diameter, spinning 800 revolutions per second, which was the speed given by Foucault to one of his gyroscopes, would advance, if allowed to roll, with the speed of 837 feet per second—nearly ten miles a minute.

There are some facts, and inferences we draw from them, with regard to motion and the geometry of space that it may be well to mention here. When we speak of the velocity of a body at a given time we mean by it that its rate is such that if continued for the whole interval of the unit of time, whether it be a second, or a minute, an hour, or any other, the body will move through the whole specified distance. A body will not need to go a mile in a minute in order to have a velocity of a mile a minute. It may not move ten feet, yet may have that or any higher velocity. This is obvious enough of course. Every one trusts arithmetical processes to lead him to correct results in velocities and time and all such familiar matters. One will say frequently, "It is six hours to New York" instead of, "It is two hundred miles to New York," and will not be misunderstood. Some persons have computed how long a time it would take to reach the sun if they were to take an express-train running at the rate of fifty miles an hour, without stopping for food or fuel; and they find it comes out nearly two hundred years,—a time of transit equivalent to five generations of men. In like manner, presuming one knows the distance to any remote point in space, the time required to get there at a given velocity one would call a simple problem in arithmetic, and it is. But there is an assumption one has to make which is rarely considered: that is, the properties of space and of time are the same everywhere, and that the geometry of the space in which we live is a geometry that holds everywhere and always: that its propositions are absolutely and irrefragably true always and everywhere. We assume, because we find them practically true on a small scale, that they are equally true on the largest scale.

Within the past fifty years the great geometers have made some very wonderful discoveries—one might say, astounding discoveries; for they tell us that we do not know that the sum of the interior angles of a plain triangle is equal to a hundred and eighty degrees, that we do not know it within ten degrees if the triangle be a very large one, such as is formed by the spaces between remote stars and the sun; furthermore, we are assured that, for all we know, and therefore for all we can reason from, space itself may be curved so that if one were to start in what we call a straight line, in any direction, and travel in it on and on he would find himself after a long time coming to his starting-point from the opposite direction; that what one would see if his sight were prolonged in any direction would be the back of his own head much magnified. Methods have been proposed for discovering if it be true or not. Some folks have called this nonsense, and have used descriptive adjectives to express their contempt for it; but none of those who have spoken thus of the new geometry are themselves mathematicians, and one is therefore left with the fair inference that they did not so well know of what they condemned as did the mathematicians who reached the conclusion.[1]

Now, we all of us trust such mathematical processes as we can ourselves handle, even when they lead us to magnitudes and distances too great for comprehension. All that one needs to know is, that the process is a legitimate one and is correctly worked out. This new geometry I have alluded to has been worked at by the best mathematicians of all the civilized nations, and they agree in the conclusions. They certainly would not do so if there were the slightest apparent reason for rejecting them; for national jealousies are too strong, and a sense of the value of truth too great, to allow

---

[1]See Appendix.

any such notions to gain currency anywhere if there were any possibilities of breaking them down.

If the space we live in and the geometric relations are only practically true upon a small scale; if we may have a kind of space of four or more dimensions, whether we now can conceive of it or not, then should one understand that spaces and distances and velocities and all computations formed upon them, though practically true, for all of our experience must not be pushed up into statements that shall embrace all things in the heavens as well as on the earth. Perhaps even the visible universe is not to be measured by our span, much less things invisible in it and beyond it.

CHAPTER IV

Energy

WHENEVER a body of matter having any motion strikes another body, it always imparts some of its motion to it, and the second body moves. The ability one body has to move another one is sometimes called its energy, and the amount of energy received is proportional to the amount of similar energy the first body possesses. A body at rest can impart no motion to another one, so it appears that the energy a body has depends upon its own amount of motion. Neither can a body impart to another one more motion than it possesses itself, and rarely or never can it do so much as that. Inasmuch as every kind of a phenomenon is the result of the transfer of some kind of motion from one body to another, one may rightly infer that to understand phenomena and their relations, one must need to know, not only the kinds of motion that are transferred, but must also know their quantitative relations, and he must therefore have some units and standards for comparison. This requires some measure for the amount of matter involved, also some measure for the motion it has. For the former it is customary to employ a weight. A certain mass of matter called a pound is adopted in England and America. Exact duplicates of its standard weights are made and preserved by each nation; so as weights become worn by usage, they may be exactly replaced. Any unit space may be adopted, as the foot, which is common. If a pound has been raised a foot, a certain amount of work has been done, which is called a *foot-pound*, and it is important to keep in mind just what it signifies. If ten pounds be raised one foot, or if one pound be raised ten feet, the same amount of work—ten foot-pounds—has been done; and with this as a starting-point, it will be easy to see how energy may be measured, for the measure of it will be the amount of work, measured in foot-pounds, it can do. It is found by experiment that if a body be left free to fall in the air, it will fall sixteen feet in a second, and its velocity at the end of the second will be thirty-two feet. If a very elastic ball weighing a pound should fall thus in the air upon an elastic pavement, it would rebound nearly to the height of sixteen feet. If it does not quite reach that height, it is because the air retards it somewhat, and some of its motion has been imparted to the pavement upon which it falls. Adding those losses to the height it did rise, and it would make the sixteen feet. Now, to raise a pound sixteen feet required sixteen foot-pounds of work; there must therefore have been sixteen foot-pounds of energy at the instant of impact. Its velocity was thirty-two feet per second. Hence a body weighing one pound, having a velocity of thirty-two feet in a second, is capable of doing sixteen foot-pounds of work. It is found also that if the same body falls for two seconds, it will fall sixty-four feet, and its velocity at the end of the second second will be sixty-four feet,—twice as great as it was for the fall of one second; but the pound weight in this case will rise under similar conditions to the height of sixty-four feet, which is four times higher than for thirty-two feet per second; so it is seen that in this case, when the velocity is doubled, the power of doing work, measured in foot-pounds, has been increased four times, and this is generally expressed by saying that the energy of a body is proportional to the square of its velocity. The particular direction in which a body moves has not been found to make any difference in this regard, so the statement is a general one. If a mass weighing two pounds were dropped, as in the first instance, it would rise no higher than if it weighed but one; but two pounds raised sixteen feet would give thirty-two foot-pounds, so the work would be proportional to the weight as well as to the square of the velocity.

The amount of matter there is in, say, a pound weight would be just the same in one place as in another; but the attraction of the earth upon it depends upon where it is. At the surface, where we measure it, it has a certain value; but at the centre of the earth it would weigh nothing. The farther it were removed from the surface of the earth upwards, the less would its weight be. At the height of a thousand miles it would be but four-fifths of a pound; at a million miles it would be but sixteen-millionths of a pound, or only about the tenth of a grain.

For that reason it has become necessary to find some measure for matter that shall be independent of position, and this has been found by dividing the weight of the body at a given place by the value of gravity at that place, and calling the quotient the *mass*; so if w represents the weight of a body at a given place, and g the value of gravity at the same place, that is, the velocity that gravity will give to a body in one second if left free to fall, then $\frac{w}{g} = m$, the mass. The distance in feet that a body will fall in a second is equal to the square of the velocity divided by twice the value of gravity, or d, the distance, $= \frac{v^2}{2g}$; and as the weight equals mg, the product of the two is

$mg \times \frac{v^2}{2g} = \frac{mv^2}{2}$, one-half the product of the mass into the square of the velocity will give the energy of a body. But it is generally more convenient to use the weight of the body instead of its mass. As $m = \frac{w}{g}$, let it be substituted for m in the expression of energy, and we shall have $\frac{wv^2}{2g} = pd$ (pressure in pounds into distance in feet), or foot-pounds, a very convenient expression to keep in mind if one has any problems in motion and energy for solution.

An example will make plain the utility of this. A body weighing ten pounds is moving with the velocity of one hundred feet in a second; how much energy has it? $\frac{wv^2}{2g} = \frac{10 \times 100^2}{64} = 1562$ foot-pounds; that is, it has energy enough to raise 1562 pounds a foot high, or ten pounds 156.2 feet high.

This is applicable to all bodies, big and little, whose weight and velocity of translation are given.

When a person who weighs one hundred and fifty pounds climbs a flight of stairs—say, to the height of ten feet—he has done $150 \times 10 = 1500$ foot-pounds of work. Whether he has gone up fast or slow makes no difference in the amount of work done; it will only make a difference in the *rate* of doing work.

Now, a horse-power is a rate of work, and is equal to 550 foot-pounds a second; and hence if the above individual climbs the stairs at the rate of four feet a second, he will be doing $4 \times 150 = 600$ foot-pounds per second, which is over a horse-power, and indicates the probability that he would not climb so fast. If any one thinks he can do it, it will be worth his while to try it.

Work can be measured on a horizontal as well as a vertical plane. Suppose the horses on a horse-car pull two hundred pounds, as indicated by a dynamometer, and the car is moved five feet in a second: the pull into the distance measures the work done; that is, $pd = 200 \times 5 = 1000$ foot-pounds, a little less than two-horse power. These illustrations are given because not every one has clear enough ideas concerning the meaning of energy and work, much less the ability to apply them to examples that may often come up. When one sees the long trail of a meteor in the sky, and remembers that its velocity may be as much as twenty or more miles per second, he will now see that it may have a good deal of energy, though its weight be but a few grains.

The energy of a pound moving twenty miles a second would equal

$$\frac{1 \times (20 \times 5280)^2}{64} = 174,240000 \text{ foot-pounds.}$$

A grain is one seven-thousandth of a pound, and its energy would therefore be but the one seven-thousandth of that quantity. $\frac{174,240000}{7000} = 24891$, which is the number of foot-pounds of work a meteor weighing one grain, at that velocity, may have: enough to raise a ton twelve feet high.

As a matter of fact, the great friction it is subject to in its path through the air heats it shortly to incandescence, and it is presently dissipated. If it were not for the air, therefore, even if we could subsist without it, mankind would be in constant danger from the flying missiles; for though they would weigh but a little, their velocity would enable them to do destructive work upon everything they struck. As there are some millions that come into the atmosphere every day, no one could be safe from them in any place.

The energy of a workingman is measured in the same way; namely, by the amount of work in foot-pounds he can do.

One of the most direct ways of knowing this for an individual is to ascertain the amount of earth or stones he can load into a cart, or the bricks he can carry up a ladder to the mason. Suppose he throws fifteen shovelfuls per minute, each one holding ten pounds, and each one is raised four feet high: then in a minute he has done $15 \times 10 \times 4 = 600$ foot-pounds of work, or 10 per second. This is rather a small quantity, only the one fifty-fifth of what a horse-power would do, and most men have been found able to do forty or fifty foot-pounds per second; still, there is a great difference in individuals in their working ability. Climbing, in general, is hard work because it is continuous lifting of one's self. One who weighs one hundred and fifty pounds, and climbs one hundred feet, has done 15000 foot-pounds of work; and if he has done it in a minute, he has spent nearly half a horse-power, which is 33000 foot-pounds a minute.

Once more: a bird in flying has to do work; and one may see how much is demanded of such birds as geese, that make long voyages through the air in the fall and spring,—sometimes for twelve hours or more continuously. As work is measured by pressure into distance, one may apply it thus. Geese are known to fly at the rate of thirty miles per hour, which is forty-four feet per second. In flying, of course, there has to be a push forward by means of their wings, not only to advance, but to maintain their elevation. Supposing that a large bird flying at this rate should have to exert a push of one pound continually: it would be expending then forty-four foot-pounds per second, nearly one-twelfth of a horsepower; and to maintain such a rate for twelve hours would imply that it had a supply of energy to start with of $44 \times 60 \times 60 \times 12 = 1{,}900800$ foot-pounds for one day's expenditure. This does not seem at all probable, and one may therefore infer that the pressure exerted when going at that rate is much less. If the pressure were but one ounce instead of a pound, the rate of work would be $\frac{44}{16} = 2.75$ foot-pounds per second, which is much more likely; but this supposes the bird to have a supply of energy of $\frac{1{,}900800}{2.75} = 700000$ foot-pounds.

In the chapter on "Chemism," the source of the energy of animals will be more particularly treated.

So far the energy involved in translatory or free-path—or, as it is more often called, mechanical—energy has been considered; but vibratory motions of matter involve energy also, and the same expression is applicable as in the first case, $\frac{wv^2}{2g}$. Here the value of the v, or the velocity, has to be determined by analyzing the motion itself. This is not simply the number of times the body vibrates, but also the extent of each individual vibration,—that is to say, the amplitude of vibration,—and the product of these two factors will give the value of v needed. So if n be the number of times the body vibrates a second, and a be the amplitude of the individual vibrations, the true velocity will be represented by an, and then the expression for the energy will be

$$\frac{wa^2n^2}{2g}.$$

For most bodies of visible magnitude the amplitude of vibration is so small a quantity that for frequencies of only a few hundred per second, the velocity, measured as a translation, is small, and therefore the energy is small, and there are few cases where it is very important to take it into account.

Suppose a vibrating body has an amplitude of the one-hundredth of an inch, and vibrates a hundred times in a second: the total distance moved through in a second would be but an inch, which would be the value of v, so the amount of energy it had would depend more largely upon the weight of the body. On the other hand, if a body is so small that its rate of vibration is exceedingly high, as was shown in the case of atoms on page 63, there might be a relatively large amount of energy involved. In the case refered to, a velocity of eighty miles a second was computed, on the supposition that the amplitude of vibration was equal to one-half the diameter of the atom; and what amount of energy is possessed by a body weighing one grain was computed. The amount in an atom with that vibratory rate and amplitude would be calculated by dividing the amount in the grain by the number of atoms in a grain. Numerically it is a very, very small quantity, and only becomes appreciable to any of our senses when vast numbers of atoms act conjointly.

There are some cases where energy is apparently expended when there is no apparent motion, as is the case when a man holds up a weight. If the weight be a heavy one, exhaustion will be the result as much as if energy was spent in any other way. This muscular work is called physiological work, and for a long time it was not understood. It is now known, however, that when a muscle is put in a state of tension, it is in longitudinal vibration a great many times a second. This may be perceived by putting the end of a finger into the ear, pressing but gently, at the same time squeezing with the rest of the hand as if grasping something tightly; a low sound will be heard, made by perhaps no more than thirty or forty vibrations per second. The muscles in a state of tension produce this. When one holds up a weight—say, a pail of water—the muscles involved yield and contract rapidly, so the weight is really raised in a vibratory way a short distance, but a great many times in a second; and the heavier the weight, the more the work done, and this too is measured in the same way as other more visible kinds. There is good reason for believing that a book resting upon a table is supported by the vibratory motions going on among the particles of the table, and therefore energy is expended to do it, and that this is supplied by the heat present in the body; that is, the temperature of the table is a little different from what it would be if it did not have any weight to support.

Walking involves the expenditure of energy in the same way. Each step requires the whole body to be raised somewhat. Suppose it be only an inch. A person weighing 150 pounds would, for each step, do $\frac{150}{12}$ foot-pounds $= 12\frac{1}{2}$. If he takes two steps per second, then each minute he does $2 \times 12\frac{1}{2} \times 60 = 1500$ foot-pounds of work. Thus one can see how physiological processes are measurable in terms of mechanical units.

The energy of a rotating body is more complicated than translational energy, because a part of the body is at rest,—the axis; and the velocity of movement at any point away from that is proportional to its distance from it. In the case of the balance-wheels of steam engines, where the most of the weight of the wheel is in the rim, the velocity of the latter would be equal to its circumference multiplied by the number of turns per second or per minute. Thus if a fly-wheel, having nearly the whole of its weight in the rim, weighs, say, a ton (2000 lbs.), is six feet in diameter, and rotates four times a second, its velocity will be 75.4 feet per second, and its energy will be $\frac{wv^2}{2g} = \frac{2000 \times 75.4^2}{64} = 177661$ foot-pounds, an amount of energy which is stored up, and may be drawn upon to prevent fluctuations in speed to which engines in workshops are liable.

If a body having rectilinear motion be left to itself in the air, it will speedily be brought to rest, for gravity will bring it to the earth whether it be moving this way or that. The air, too, will retard its motion, and would ultimately bring it to rest if nothing else did, as it would either of the other kinds of motion. If, however, one could contrive to give to a body above the atmosphere a sufficient velocity in a tangential direction, the body would become a satellite, and revolve round the earth. The curvature of the earth is about eight inches to the mile, and such a body would then need to move a mile in a horizontal direction in the same time it falls eight inches in order that it should continue to go about the earth. As it takes about two-tenths of a second to fall this distance, its velocity would need to be five miles a second to prevent it from falling to the earth; this velocity would carry it quite round the earth in a little less than an hour and a half.

Thus it is seen that, in order that matter should possess energy, it must have motion of some kind; indeed, that energy has two factors, mass and motion. When either of these is zero, there is no energy. This is a consideration of great importance both in a scientific sense and a philosophical one. One may often hear it said and read it in carefully written books that matter and energy are the two realities or physical things in the universe, and energy is spoken of as if it were an entity, or something that might exist though there were no substance to move. If energy be a product, and motion be one of the factors, then in the absence of this there is no energy. This perhaps will be seen still clearer after considering what are called the laws of motion, which were first formulated by Newton, and which, in conjunction with the law of gravitation, were the fundamental principles that enabled him to produce the "Principia," which is what to-day we would call a treatise on mechanics.

Of course, the science of mechanics is applicable to motions of matter of any magnitude and in any place; and Newton chose to follow out his newly discovered principles into astronomy to the largest extent, and it remained for later generations to employ the same principles in other directions, largely molecular and atomic.

The first law of motion is, that whether a body be in a state of rest or of motion, it will remain in that state of rest or motion until compelled by the action of some other body upon it to change its state. This is sometimes expressed by saying that all matter has *inertia,* or an inability to move or change its direction or velocity if it has motion. This appears to be experimentally true of all bodies whose magnitude and state we can see. But it may very well be doubted if the ordinary conception of the inertness of matter be true. Many of the facts of chemistry indicate that matter in its atomic form is not altogether so helpless as it has been supposed to be. A stone may lie in the road for an indefinite time and no one would suspect it possessed any energy to do anything, and so of any other kind of matter. Here is a piece of charcoal. Has it inertness in any extreme sense of that word? Here is some sulphur and some nitrate of potash; they, too, will lie as quiescent as the coal and as long. Pulverize them and mix them together, and we have powder the energy of which would wreck a building. The products of the explosion are gaseous mostly, and the carbon, the sulphur, and the nitrate of potash have vanished as such, and have entered suddenly into new combinations; they have developed also a large amount of heat, while at the beginning their temperature was that of other bodies around them. This source of energy must have been resident in the atoms; and if it is perceived that for a body to have energy it is necessary for it to have motion of some sort, it will be apparent that the material itself must have possessed a large amount of motion, even when it appeared to be at rest. If one thinks that the law of inertia might still apply to atoms, and that they cannot individually move except as they are acted upon by other atoms, and even then only as much as by the measure of the motion thus imparted, he had better figure out to himself the energy of such explosions per molecule, and see if anything initially done will account for it.

When the mechanism of a clock is running, the motion may be traced to a falling weight, and the work done is measured by the product of the weights into the distance it falls as the clock runs down; but in the case of the powder, though the amount of energy developed by the explosion is definite, it is not measured by the work done in pulverizing and mixing and igniting it. The case is much more nearly analogous to that of a sleeping man. While asleep he would neither move nor stop moving unless some other agency acted upon him, any more than would a stone or other mass of matter; and in that sense he would be inert, yet no one would think of calling a sleeping man inert, except in a very loose sense.

Furthermore, there is an experimental analogy that may help one to see a little deeper into this. Every one knows what is meant by the "sleep" of a spinning top. It appears to be absolutely at rest, and may not even hum; but touch it, and the effect upon it will be out of all proportion to the slightness of the touch.

It has been observed as a property of vortex rings that they have a tendency to move forward in the direction of their axes, and when prevented from going forward they press upon the body that arrests them. If they be brought to rest, and then the barrier be removed, *they, of their own accord,* start on in the same direction as if pushed from behind. Such a body cannot be said to be inert without modifying the common meaning of the word.

This is not alluded to here as proving anything; but inasmuch as the vortex-ring theory of matter has a good probability in its favor, this property I have mentioned helps one to understand how the atoms might be other than inert, and yet large bodies of them together exhibit that property with the rigorousness our observations upon such bodies demonstrate. Suppose each atom had the ability to move forward of its own impulse when not acted on by any other atom. If there were a million atoms joined together, no matter how, provided they were promiscuously faced, they would mutually neutralize each other's ability to move in any direction, and the resultant of the whole would be that passivity which we call inertness.

We may by and by see that there may be still other good reasons for thinking matter not to be so passive as it has been often assumed to be.

The second law of motion is, when two or more bodies act upon a third body, the effect of each is the same as if it alone acted, and the combined effect is called the resultant; and the third law is, that action and reaction are always equal and opposite in direction. This third condition of action, or the relation of motions in two bodies, is of a high degree of philosophical importance, perhaps not more so than the others, but of so much that it is worth while to attend to it more particularly than to the second law. If a rope be tied to the wall and one pulls upon it so as to make it taut, the wall pulls back in the opposite direction as much as the arm pulls forward. A spring-balance attached to the wall would indicate the strength of the pull, the pull of the arm representing the action, and measured by the muscular vibration, as already described, and the pull of the wall representing the reaction, and equal to the action in quantity and maintained by molecular vibration. Imagine the action of the arm to be steadily increasing in quantity: the reaction of the wall would correspondingly increase until the molecular tension could no longer be increased, and either the rope would break, the hook be pulled out from the wall, or the wall itself be broken away; but in no case could the action exceed the reaction or *vice versa.* Now, if the amount of matter in the arm were a constant quantity, as well as that of the rope, the hook, and the wall, then it would follow that all the physical changes noted in either the one or the other, so far as energy is concerned, must be due to the motions involved on either side. And if action and reaction be equal, and the quantity of matter be uniform, then the amount of motion involved must be equal on the two sides. If a body in motion strikes another body, and the second one is set in motion, the amount of motion in the two will be just equal to the amount of motion in the first. The amount of motion gained by one body is just equal to that lost by the other, and there has been simply an exchange of motions, one having gained, the other lost; the one that gained being the one that had less, and the one that lost having had more, than the other one. In books of physics it is customary to speak of the amount of motion a body has as its *momentum;* and it may be measured by multiplying the mass of the body by its velocity, and oftentimes one may read that in the physical exchanges that are all the time happening in matter the momentum is conserved; that is to say, is neither increased nor diminished. Seeing, therefore, that the amount of matter is a constant quantity, and the momentum a constant quantity, it follows that the amount of motion is constant. Motion is conserved as well as matter. If the amount of matter in the universe be constant, then, according to this statement, the amount of motion must be constant, and the amount of energy constant also.

It is generally agreed that this statement concerning energy is true, and one hears often about the law of the conservation of energy. It seems to be less clearly recognized that the third law of motion implies the conservation of motion, provided matter is itself a constant quantity. But there is another condition of things that is as uniform as any other condition of things in nature that has not been recognized as a law, and yet it deserves to be perhaps as much or more than most others, since, in our experience, it is never known to vary; it is this: Wherever there is an interchange of motions between two bodies, the transfer is always from the one having more to the one having less. As exchange of motions implies transfer of energy, it follows that all transfers of energy of any given kind are from bodies having more to those having less.

Cause and effect are always determined by such a disposition of things, though not every one has apparently seen that questions involving what they please to call causes and effects presume a kind of antecedent and consequent that always work both ways at the same time, for there is no such thing as an isolated phenomenon. If everything takes place so and so because there is an exchange of motion going on, then this thing that now moves faster than it did has been acted upon by a body that had more motion in this direction than the former one had, and it has imparted some of its motion at the expense of its own energy. If one inquires what caused the increased velocity to this body, it may be said it was caused by the impact with another body. In like manner one may inquire what caused the slowing-up motions of the second body, and the answer still must be, the same impact with the first body. So, for every phenomenon there is a corresponding and complementary phenomenon, which it is just as appropriate to consider as a cause as it is the first, and either element is just as much a cause as the other, and in each and every case all there is involved are exchanges in the amount and kinds of motion in matter.

There remains now the consideration of a topic

which those who have studied physical subjects only a little must be more or less familiar with. The term "potential energy" has been much employed within the last twenty years to express a certain condition of matter that renders it a source of energy when no motion is supposed to be involved: thus, where a weight is raised, like that of a clock, or of a stone raised to the roof of a house. By falling, either of them can be made to do work; but so long as they remain raised and are apparently quiescent, their stock of energy is measured by their weight into their height, i.e., foot-pounds; and this is said to be *potential energy*. Examples of this sort are numerous. The wound-up spring of a clock or watch, a bent bow, compressed air or steam, powder, nitro-glycerine, and the like explosives, coal, wood, and other kinds of fuel, are all varieties of so-called potential energy. Let it be remembered that we have in natural phenomena matter and ether and space and time and motion. If matter and ether be substances, then the product of one into the other would signify nothing; it would be physical nonsense. So likewise would be the product of matter into space or time; and yet if matter is to be possessed of energy, and motion is *not* one of the factors, then either space or time must be, and no one can imagine how energy can in any way depend upon time as a factor, and there is no degree of probability that it is or can be so; and hence, though we had no hint of how it might be, one would need to avow his belief that in some way motion was involved in every case where physical energy was involved, for in any case where it had been hitherto possible to trace it, it had been found to be present as a factor in precisely the same relations as in all other known cases, and hence he would avow a disbelief in the existence of potential energy in any other than a loose sense for a condition where the character of the motion involved was obscure. This would imply that all energy is kinetic, whether the character of the motion was determined or not. This view is now held by those who have taken the pains to think out the necessary relations that are involved in this subject.

In the last edition of the "Encyclopædia Britannica," Professor Tait, who contributed the article on "Mechanics," says, "Now, it is impossible to conceive of a truly dormant form of energy whose magnitude should depend in any way on the unit of time; and we are therefore forced to the conclusion that potential energy, like kinetic energy, depends in some unimagined way upon motion;" also, "The conclusion which appears inevitable is that whatever matter may be, the other reality in the physical universe which is never found unassociated with matter depends in all its widely varied forms upon motion of matter;" and in another place, "Potential energy must in some way depend upon motion."

It was pointed out (on p. 67) that what was called physiological work is now known to depend upon the vibratory state of muscles in a state of tension. Before that explanation was known, one might have called such, potential energy, if it had not been for the sense of fatigue felt by one who was doing such physiological work that forbade him to assume that actual energy was not employed to maintain such a stress; and when it becomes evident, as it has, that one cannot press upon a table, or pull upon a rope, or bring about in any way a push or a strain upon matter, without varying the temperature of the body, it is no longer difficult to understand that all changes of that sort upon matter result in atomic and molecular stresses, for they are placed in abnormal positions as well as stretched muscles, and their energy is spent in a similar manner. There is a curious phenomenon exhibited by all bodies that are made to do atomic and molecular work for a considerable time. They become exhausted, like living things, and require rest to recover their properties. Thus, a tuning-fork, if kept artificially vibrating for some time, will stop almost instantly when the driving force is stopped, though at the outset it would continue to vibrate for a minute or more when left to itself. This is caused by what is called the fatigue of elasticity: the body loses some degree of its elasticity, and requires time to recover it. I have called the phenomenon curious. Perhaps it is no more so than any other phenomenon manifested by matter; but it is so similar to what is so characteristic of living things, that it almost excites one's sympathy. One can have compassion for an overworked and exhausted horse, but an overworked tuning-fork! The expression would seem to be wholly inapplicable, but the fact is as stated. The only difference between the cases is, one has nerves, and becomes conscious of the exhaustion, the other not.

So far, both motion and energy have been considered as related to matter, and matter as defined in the first chapter, as distinguished from the ether, though immersed in it, and can by no means be isolated from it; but energy exists in the ether as well, as we are assured by many phenomena. That light requires about eight minutes to come to us from the sun has been proved in numerous ways. When it gets to the earth it is found to be able to impart energy to the matter it falls upon: it may heat it and affect it in other ways that are measurable, so energy gets to us from the sun, and is eight minutes in transit in the ether. If we do not call ether matter, and it has been shown that there are good reasons for not doing so, then it follows that energy exists outside of matter, and it is a proper line of inquiry to learn what shape the energy exists in, and what mechanical conceptions are appropriate when thinking about it. In matter one may isolate motions of various sorts. A mass of matter, say, like a baseball, may have translatory motion: it may vibrate or it may spin. In each case one may contemplate the kind of motion, and compute the energy involved in the movement, and this is true for atoms as well as larger masses; but when the substance is not made up of discrete parts, but is absolutely homogeneous with no interstices, and apparently incapable of changing either its position or its form, as there is good reason for thinking to be the case with the ether, it becomes much more difficult to picture to one's self just what is happening when motion of any sort is involved. As has already been said, we know that light consists of waves, measurable quantities, and we know how much energy reaches the earth from the sun and falls upon a square mile or square foot. There have been several estimates of this quantity, and it is found to be equal to about one hundred and thirty foot-pounds per second for each square foot section of sunshine. This signifies, of course, that that is the amount of energy in a column of ether one foot square and a hundred and eighty-six thousand miles long, for that is the amount that arrives per second. So one may calculate the amount of energy there is in a cubic mile of sunlight to be about twelve thousand foot-pounds, and also that the amount given out by the sun in a second is about four millions of foot-pounds, or nearly seven thousand horse-power for each square foot of the sun's surface. All of this energy is handed over to the ether, which distributes it in all directions as undulatory movements which we call light. Such wave motions do not exhibit anything like what we call momentum as waves in water or air always do, and they are therefore in striking contrast with waves in matter. Moreover, being waves, having the amplitude at right angles to the direction of propagation, they must be compounded of two motions,—a rectilinear and a vibratory one,—and not a simple one such as a particle of matter may have.

The ether is capable of being affected by other motions of matter than simply the vibratory one of atoms and molecules.

Whenever an electro-magnet is made, it reacts upon the ether in such a way as to affect other matter that chances to be in the range of ether so affected. It appears as if the ether were thrown into a state of stress which it retains so long as the magnet retains its property; and this condition extends to an indefinite distance in all directions. If such an electro-magnet is made and unmade by opening and closing an electric current in its coils, there will be formed a set of electro-magnetic waves in the ether which will travel outwards from the magnet in a manner similar to light-waves, only they will have an enormous wave length. If the circuit be closed but once a second, the waves will be a hundred and eighty-six thousand miles long; for a wave in the ether travels in it with a velocity that depends solely upon the property of the ether to transmit disturbances, and not at all upon the source of the disturbance. That such an electro-magnetic wave possesses energy, and can do work, one may satisfy himself by observing the motions produced by them upon magnetic needles within the affected space.

In like manner an electrified body puts the ether into a different kind of a stress from the magnet; and when this is done periodically, as it may be by an induction coil, and in other ways, electrostatic waves are set up, and these too travel with the speed of light, and are capable of affecting matter to a great distance, thus showing that the ether may possess energy in an electro-static form, as distinguished from the electro-magnetic and light. There are some physicists who think these last two to be identical, and the reasons for their opinion will be given in a subsequent place.

It only remains to point out that whatever the nature of gravity may be, there can be very little doubt that the ether is intimately concerned in it, as Sir Isaac Newton supposed was the case. But if it is, and ether is the agency by which one mass of matter is able to affect another mass, then ether is in a state of stress produced by the atoms of matter all the time, and therefore in some way gravitative energy is lodged in it. As the ether is so universal in its extension, one cannot but see that it is a storehouse of an almost unlimited amount of energy of many kinds; so that if every particle of matter were instantly annihilated, there would still be a universe filled with energy, though it might not be serviceable, because lacking the conditions for transformation into useful forms. This may be said to be one of the functions of matter—the transformation of the energy it gets from the ether.

CHAPTER V

Gravitation

THAT all bodies will fall towards the earth if raised above its surface and left unsupported everybody knows and must always have known, for it is a fact thrust into everybody's notice constantly and as long as he lives. Also that bodies resting upon the earth require energy to be spent in order to raise them from it is equally well known. Thus all bodies act as if they were attracted by the earth, and the weight of a body is the measure of the attraction of the earth upon it.

One not unfrequently comes across statements by authors implying that Newton was the discoverer of this attraction which is called gravitation. This is a mistake: not only was this idea common in Newton's day, but the word itself was in extensive use. Kepler had affirmed that the sun attracted the earth and the planets, and Galileo had busied himself very much with the study of attraction of the earth upon bodies. The problem that Newton had before him was not as to the existence of gravitative action, but what was its law of operation and the limits of it, if it had any limits. The familiar story of the fall of the apple leading to the great discovery is generally believed to be mythical; at any rate, other facts well authenticated do not accord with that story. When he was twenty-three years old he undertook to apply the law as we now have it, to the moon, using the size of the earth and the moon's distance from it, as they were then best known. The result satisfied him that his surmise could not be the law, if the measure of the earth then had was accurate. This was in 1666. In 1683 he learned of some new measures recently made of the magnitude of the earth, indicating it to be larger than had been supposed. Then, with the new measures for data, he made a new computation. It was then, when he saw that the results were to prove his conjecture, and he perceived the immense importance of the discovery, that he handed over the unfinished work to an amanuensis, because he was too much agitated to complete it. If the discovery was made when he first thought of putting the idea to the test, it is strange that his emotional excitement should have been postponed for seventeen years. Evidently it was at the latter date when he thought he had made the discovery. It was the *law* of gravitation that Newton discovered, and that it was universal. Every particle of matter attracts every other particle; and the strength of this attraction varies as the mass of each, and inversely as the square of the distance between them. Thus, if at the surface of the earth gravitation gives a weight of one pound to a body, at the distance of ten radii of the earth = 40000 miles, the weight would be $\frac{1}{10^2}$, one-hundredth of a pound, and at the distance of the moon, or sixty radii of the earth, the body would weigh but $\frac{1}{60^2}$=one thirty-six hundredth of a pound, and would fall towards the earth in a second but $\frac{1}{3600}$ of the distance it would fall at the surface of the earth, where it is about sixteen feet. One thirty-six hundredth of sixteen feet is about the one $\frac{1}{224}$ of a foot, which is therefore the departure from a straight line the body at the distance of the moon must make per second to move round the earth. The mutual attraction of these bodies at that distance is sufficient to produce this amount of deflection, and hence accounts for the rotation at that distance. When the same mathematical relation is applied to the planets, comets, and meteors that revolve about the sun, it is found to be applicable to every one of them; and in the depths of space in every direction are to be seen multitudes of stars revolving about each other in similar manner, and hence it is concluded that gravitation is a universal property, and the law is applicable throughout the universe.

There are other kinds of attraction that matter exhibits, such as electric and magnetic, that follow a part of the above law, but do not the other part. The law regarding the distance is true for electrified bodies, but the mass of the bodies does not enter as a controlling condition. So it appears that the variability of attraction with the distance is a geometrical condition, and depends upon the property of space, and is not peculiar to any physical phenomenon. Sound, light, heat, electricity, magnetism, as well as gravitation, exhibit the property, as do circles and spheres. The peculiar thing about gravitative attraction is that it depends upon the masses of the attracting bodies, and is not modified in the slightest degree by the interposition of any substance of any magnitude between the attracting particles or masses. In this particular it is strikingly unlike magnetic attraction. If, for instance, a piece of iron is brought between two magnets that at a distance are attracting each other, the strength of their action upon each other is decidedly less. The strength of the attraction of the sun is just as great upon a particle in the centre of the earth as for any similar particle at an equal distance that is not shielded.

There have been numerous attempts in the past

to account for gravitation. It has been imagined that space was full of particles swiftly moving in every direction that produced a pressure upon all bodies by their impact; that each body shielded other bodies in a measure, and hence the pressure produced upon the adjacent sides would be less than elsewhere, and, as a consequence, each body would be pushed in the direction of an adjacent body. But a push represents expended energy, and this would imply that the moving particles must be losing energy at the expense of their velocity; and as no such particles are known, and if there were, their velocity would have to be so much greater than that of light, there is no degree of probability to be allowed for the idea. The effect of vibrations upon the ether has been a very common manner of attempting to explain gravitation. It has been observed that if light bodies are brought near to a vibrating body like a tuning-fork, they are apparently attracted by it so long as the vibratory motion continues; and the action is explained by the rarefaction produced by the vibratory motion, which reduces the pressure in the space about the body, so when another body is brought near the pressure is greater on the remote side than it is on the side adjacent, and thus the body is pushed towards the one vibrating. It is known that all the atoms of all bodies are in a state of vibration at all temperatures; and hence it was inferred that the pressure of the ether must be reduced next to their surface, so that between two atoms or molecules the pressure must be less than external to them, and hence the pressure of the ether will crowd them together. This idea has been worked out by a large number of persons in different countries. There are two fatal objections to this hypothesis: First, it would make the attraction of gravitation dependent upon their temperature, and there is no evidence to show that temperature makes any difference; and second, that the velocity of gravitative action is the same as that of light. There is an abundance of astronomical evidence, that if it has any velocity at all it must vastly exceed that of light. If it were as much as a million times greater, astronomical phenomena would exhibit it plainly.

Seeing that every particle of matter in the universe, affects every other particle in a certain and definite way, no matter what the distance between them, there must be either the possibility that a body can act upon another one at a distance without any medium between them,—which is called action at a distance,—or there must be a medium which is first affected by the bodies, and which in turn reacts upon other bodies in it. What Sir Isaac Newton thought of these contingencies was cited in a former chapter (see p. 36). It is now generally felt to be not only essential for consistent mechanical thinking, but that in some way the ether which is known to exist must have some essential part in the phenomenon. It has been the subject of curious speculation why Newton should so strongly state his belief in the existence of a medium for the propagation of physical conditions, and yet in his work on light he should adopt the corpuscular theory—that light consisted of emanations, which was a practical denial of the hypothesis of the ether. The explanation of the anomaly is probably in the fact, that he could treat in his mathematical way the ideal corpuscles, while he could not so treat the ether hypothesis of waves. His work was developed with ideas he could handle; and the outcome of it was that the science of light was retarded by his misconceptions for a hundred years, for every one now who knows anything about it knows that Newton's hypothesis was a wrong one. There are some persons who would curb the imaginations of others in physical things by quoting Newton's dictum, "Hypotheses I do not touch," but they omit to mention that Newton's work on optics was altogether based upon a hypothesis that has wholly broken down. Every one of the explanations he gave of such phenomena is worthless, and no one gives attention to them except for their historic relations to the science.

It has been thus in other lines. A symbolic representation of things such as offers the possibility of mathematical treatment has been seized and worked out to great length, when the actual phenomena pretended to be treated gave no countenance to the conceptions. Such has been the case in electricity and magnetism and heat. The mathematicians fought Ohm's, Faraday's, and Joule's mechanical conceptions until death stopped them.

It is certainly true that all physical phenomena are subject to strictly mathematical conditions, and mathematical processes are unassailable in themselves. The trouble arises from the data employed. Most phenomena are so highly complex that one can never be quite sure he is dealing with all the factors until experiment proves it. So that experiment is rather a criterion of mathematical conclusions and must lead the way. Mathematics is a deductive science, yet the number of physical facts or phenomena that have been discovered by its aid is so small that they may almost be left out of the count. There is the discovery of the planet Neptune, that has been spoken of as a triumph of mathematical science, yet one of the most competent mathematicians that ever lived—Professor Peirce of Harvard—declared that it was only a lucky find, for the computations would apply just as well to a planet 180° from it. The conical refraction of light is another one. Altogether they make but a small figure and are unimportant. The law of gravitation was discovered by trial, and although its importance is second to none other yet discovered, it happens that it is one of the very simplest and least complicated with other laws we know of; but an explanation of how it can act thus, or why it exists at all, or what its antecedents are if it has any, these are questions that are matters for the guessers, like Kepler, who kept guessing until he guessed right, and so discovered what are known as his laws. Meanwhile definite mechanical conceptions of what the phenomena to be explained are like may be helpful to those interested in them.

Suppose two bodies, *A* and *B*, a certain distance apart, and they so react upon each other that they tend to mutually approach each other. Given a medium, ether, can one imagine stresses set up by either body in the ether that will be capable of affecting the other?

Imagine a large space like a room occupied by glass of uniform texture and properties throughout. If relieved of gravitational property, the cohesion of all its parts shows that every particle is in some sort of stress, no matter what the origin of that may be. Now, suppose there could suddenly be created somewhere near the middle of the glass a bullet or a marble. It would displace so much glass as would be equal to its own volume, and the result of that would be that the glass about it would be subject to a new stress, which would be greatest, at the surface of contact of the marble, and would be less as that surface is receded from inversely proportional to the square of the distance of the point of observation. If the glass be imagined to be indefinitely great in magnitude, then the stress would extend in every direction through the whole extent of it, and at any assignable point would still be in accordance with the inverse law, diminishing outward. Imagine now another similar marble to be created at the distance of a foot from the first. Inasmuch as it displaces so much glass it will set up a new stress in the latter, and this stress must also be transmitted throughout the whole mass as in the first instance. Now, here will be two independent stresses overlapping; and on account of the nature of the stress, it will be greater between the marbles than it will be anywhere else, because there the sum of the stresses will be at a maximum. If one can now for the moment imagine that the glass was of such constitution as to permit a motion to the marbles, in any direction, when there was a stress tending to move them, it would be obvious that the marbles would separate from each other as the medium, the glass, was under greater tension between them than in any other direction.[1] And if the glass thus mobile was indefinite in extent and without friction, the two marbles would continue to separate indefinitely. The energy making them thus to move comes directly from the medium, which in turn got it from the bodies themselves when they were thrust into it, no matter how. Such a phenomenon as separation in a manner like the above is exactly opposite in character to that of gravitation, but it points at once to a consideration of the condition necessary to be similar. It was the forcing of new material into space already occupied with other material that developed the stress and led to the above results. It will be necessary to find a way to develop a stress *towards* a point instead of away from it.

Suppose, then, that instead of having a created something imbedded in it, a cavity of equal volume to the marble should be produced in its place. As part of the material of the medium has been annihilated, there will now be a less stress at its bounding surface than there was when it was occupied with material, and the direction of the stress will now be towards the cavity. That is, the stress will be less there than

[1]Such bodies might be said to have *negative weight*.

anywhere else in the glass; and this, too, if measured, will be found distributed like the other, inversely as the square of the distance from the origin. Let another similar cavity be produced in the neighborhood of the first, and the two stresses will overlap, and there will be less between them than in any other direction. Let us imagine now that the glass was mobile enough to permit the movement of either of these cavities in any direction towards which there was any pressure, and they would approach each other because pushed by the stress in the glass more towards each other than in any other direction. If one of these cavities were larger than the other, one would expect that the corresponding stress would be greater, and so there would be a stress that for direction and the resultant movement would correspond with what is observed in the phenomena of gravitation.

But such a conception as that of a vacuum as constituting what we call the atoms of matter has no mechanical validity at all. Atoms have not only volume, they have mass, and that requires energy to displace. One cannot imagine that the displacement of an absolute vacuum, if such a thing could be done, would require any energy, for there would be no mass to move.

Suppose, however,—instead of imagining, as was

done, the entire volume of the marble to be destroyed,—that in some way the volume of the glass marble had suddenly been reduced, no matter how, and that the diminished volume was retained,—the material had been condensed. This would bring about the same relative condition of stress to the condensed portion, so that there would be less adjacent to it than elsewhere, the measure of it being the actual amount of condensation represented in the body. What would be true of one would be true of others,—an indefinite number,—and no number of such stresses would in any manner interfere with or neutralize that of others. At any point of the space filled with such glass each such condensation would have produced its effect at the outset, and if the glass were practically limitless in extent this relationship would be maintained so long as the reduced volumes remained constant.

So far has been considered a condition of things somewhat analogous to gravitation; and to apply it one needs to imagine the ether to be substituted for the glass and the atoms of matter for the imagined condensation, and also that the two, the ether and atom, are capable of mutual reaction.

There have been some physicists who have imagined that the atoms of matter were condensations in the ether, but I am not aware that any very satisfactory reasons have been given for thinking so. That in itself would be no reason for rejecting the idea in the absence of a better and more consistent one. For scientific purposes a poor hypothesis is better than none at all.

A very large amount of scientific work has been done by employing hypotheses that are now known to be wrong. A working hypothesis is needful. If it be wrong, one will by and by find it out and be able to amend it, or replace it by a better. If it be right, it will be vindicated, and will justify itself, and be generally adopted.

Until we know more definitely than is now known what the constitution of matter really is, one can only guess and try; and among the multitude of interested workers in all civilized countries there will be some who will guess right to the advantage of all.

If, then, one adopts the vortex ring theory of matter, and endeavors to trace the mechanical conditions that might obtain with such kind of atoms, he would be led to inquire whether a vortex ring does or does not exhibit any evidence of condensation in the material that is in rotation; that is, does the material of the ring occupy the same space while it is in rotation as it does when not?

There are several phenomena that seem to show that it occupies less space. The reduction of pressure in its neighborhood shows a rarefaction there, and the mutual approach of such rings and of other light bodies in their neighborhood indicates the same thing. If one rotates a disk rapidly, any light bodies in front of it will tend to approach it even from a distance of several inches. If a dozen disks five or six inches in diameter are set loosely an inch apart upon a spindle a foot long, so that they may be rotated fast, yet left free to move longitudinally upon the spindle, they will all crowd up close together as the pressure is less between them than outside. If one can imagine the spindle to be flexible and the ends brought opposite each other while rotating, it will be seen that the ends would exhibit an apparent attraction for each other, and, if free to approach, would close up, thus making a vortex ring with the sections of disks. If the axis of the disks were shrinkable, the whole thing would contract to a minimum size that would be determined by the rapidity of the rotary movement, in which case not only would it be plain why the ring form was maintained, but why the diameter of the ring as a whole should shrink. So long as it rotated it would keep up a stress in the air about it. So far as the experimental evidence goes, it appears that a vortex ring in the air exhibits the phenomenon in question. There is no doubt at all that two vortex rings in the air attract each other, for they will mutually approach if free to do so, and the explanation is plain that there is reduced pressure between them; in other words, the characteristic motion of the ring reduces the air pressure about it, so that another body within that field is pushed towards the place where the pressure is least. The reduction of the pressure about any ring must evidently depend upon the amount of material embodied in it, and more especially the degree of rotation which it has. A small, thin but rapidly rotating ring might produce as great a rarefaction about it as a much larger one with less velocity, hence there is something about it that corresponds to what is called mass. It is not *simply* an amount of material, but the *energy* the material has, which gives it its characteristic properties.

Analogy must not be mistaken for identity. There is so great a difference between the properties of the air and other gases and those of the ether that one cannot affirm that what holds true of one must hold true of the other; yet that is what is generally done by such persons as those who try to show the properties of the ether to be identical with those of matter.

We know what conditions are necessary in order that a ring should be formed in the air, and one of them is that there must be gaseous friction. If that were not the case a ring could not be formed. If the ether be the frictionless medium it is generally supposed to be, one would not know how to make a vortex ring in it. On the other hand, the reason a ring in the air is so soon destroyed is because of friction; and hence if one were made in some unimagined way in the ether it would continue to exist indefinitely, but how it could act at all upon the ether surrounding it would be a mechanical puzzle, and that is the present state of the case. The puzzle is no greater with the conception of a vortex ring than if the atom were made up in some other way, and therefore that objection is not peculiar to this hypothesis. It has been confessedly a puzzle to see how the vibratory motions of atoms and molecules could set up transverse waves in the ether if the ether be without friction; nevertheless, they do set up such waves. A common objection to all attempts that have been made to account for gravitation by means of the motions of the atoms themselves is that it not only requires a constant expenditure of energy, but that the velocity of transmission must be so much greater than that of light. Light is transverse vibratory movement. A direct longitudinal wave may be much swifter than the other. A pull upon a taut rope will travel much faster in it than will a wave produced by a transverse movement of the hand.

It is not to be understood that what is presented here is given as a proof that gravitation is but a simple mechanical condition of things. It is probable that every one who thinks about it believes that its explanation is purely mechanical. Some perhaps are pessimistic, and doubt that man will ever be able to understand its mysteries, but pessimists are not discoverers. They frequently so chill the air about them that more hopeful ones, who are not persuaded that the end has yet been reached, are sometimes deterred from venturing into fields where they have to pass such self-constituted gate-keepers.

There are few physical problems of any generality and complexity that are abruptly and completely solved by one person. Tentative steps must be taken, and much labor is oftentimes spent upon ideas that by and by are proved to be worthless. A good deal of the work done by Laplace upon the Nebula theory was of that sort; yet all astronomers hold the Nebula theory in some form: what the exact process was, if solely mechanical, may be interesting, but not very important from a philosophical standpoint.

So one may hold that gravitation is a mechanical action, and in some way explainable on mechanical principles, even if he does not see how at all.

This chapter may help some to see not only what

the character of the problem is, but what factors are present, and how somewhat similar phenomena may be reproduced at will; but the radical distinction that exists between the ether and matter must always be kept in mind.

CHAPTER VI

Heat

HEAT and cold are two words we apply to contrasted sensations, either of which may imply comfort or discomfort; and what is meant by either word in a given case depends altogether upon what the sensation is compared with. Thus, one would speak of a day when the thermometer indicated one hundred degrees in the shade as being a hot day, while if his cup of coffee had the same temperature it would be called cold; so the terms imply only roughly some departure from a standard of comfort. To obtain more definite knowledge of that physical condition which gives us the sensation we call heat, it is necessary to attend to its origin and its effects upon other bodies.

I. MECHANICAL ORIGIN.

When a blacksmith hammers a small piece of iron, like a nail, upon his anvil, it becomes too hot to hold, and it even may be made to glow, red-hot, by the repeated blows of the hammer. If a bullet be shot against a target and be quickly picked up, it is found to be hot; and in general the impact of any two bodies always results in heating both of them. In the above cases both the hammer and the target are heated, but on account of their size the degree of heat is not so noticeable as it is with the smaller bodies.

In like manner if the knuckles be rubbed briskly upon one's sleeve, the sensation of heat becomes unbearable in a very brief time. The friction of the surfaces develops the heat, as may be learned by taking a button or some similar object, and in the same brisk manner rub it on the sleeve or other convenient surface, and it will get too hot to be safely touched against the skin. On a larger scale the brakes upon railroad-cars exhibit the same quality when they have been applied for a few seconds. The sparks that may be seen flying from them in the dark is testimony to the same thing; while the car-wheel boxes are often so heated by the constant friction when the lubricating oil is wanting, that the cotton waste takes fire, and even locomotives may be delayed by their hot journals. This source of heat is so common that instances may be cited indefinitely. It is universally true that the friction of one body moving in contact with another heats them both, and the heat developed depends upon the pressure and the velocity of the moving surfaces. It is true not only for solids, but for liquids and gases as well, and the friction of solids moving in either liquids or gases. An extreme case of the latter kind is illustrated by the shining trail of a meteor when it enters the atmosphere. Its velocity is very great—twenty or thirty miles a second—and the friction of the air is so great on account of the high speed that it renders the surface of the meteorite red-hot, and some of its molecules are ground off as they would be if it were held against a swift turning emery-wheel that scatters the sparks in the air. The luminous trail consists of these heated particles. If the body is not large, and most meteors are quite small, they may be entirely ground to powder and dissipated before they can reach the earth. Most meteors in this way rarely pass through more than fifty or sixty miles of our atmosphere before this happens.

Another mechanical source of heat is compression. Let a bullet be hard squeezed in a vise, or in any other way, and it is found that its heat is perceptibly increased. Small differences of this sort may be easily detected by the use of the thermopile and galvanometer.

The rubbed button or pounded or squeezed bullet placed upon the face of the thermopile shows the presence of an amount of heat which the sense of heat would

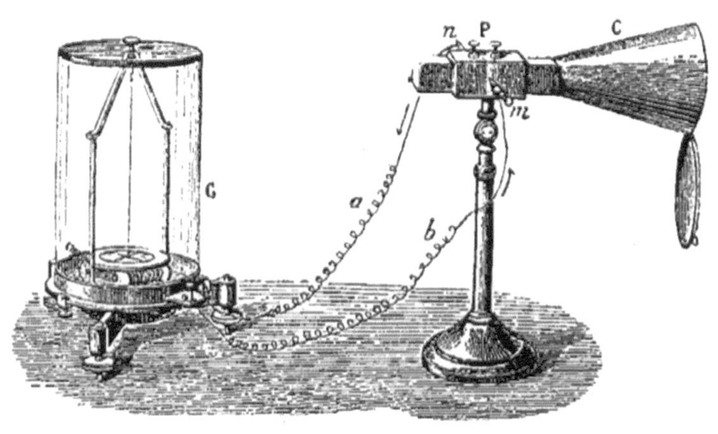

DIAG. 4.

DIAG. 5.

never detect. Gases exhibit the heating effect through pressure in a high degree. Before the invention of friction matches, which are themselves good examples of the production of heat by friction, metallic tubes, closed at one end with a tight-fitting plunger to be worked by hand, were in common use for lighting fires. A bit of punky wood was fixed to the end of the plunger, and the latter was then quickly driven to the bottom of the tube. The air was compressed to so great an extent that the heat developed became sufficient to ignite the punk. The same heating effect of compression may be shown by the thermopile and galvanometer by compressing the air with an air-condenser, and permitting the air thus condensed to strike on its exit upon the face of the pile.

Thus impact, or *sudden* stopping of mechanical motion, friction, or the *gradual* stopping of mechanical motion and condensation, or compelling molecules to occupy less space, all of them of a purely mechanical nature, result invariably in heating the matter that is subject to the action.

II. CHEMICAL ORIGIN.

The heat that results from the combustion of fuels of all sorts is due to the chemical changes that take place. When coal burns, its substance, carbon, is entering into combination with the oxygen of the air, and a new chemical product is formed called carbon dioxide, which is a gas; and the change is accompanied by the production of a large amount of heat, which we utilize for our comfort or for the various arts that depend upon heat as an agent. Wood, alcohol, the various oils,—everything capable of burning, and which may be called fuels—are, in the process of burning, undergoing what is called oxidation, in which new chemical compounds are formed and which are nearly all gaseous. Thus the products of the combustion of wood, alcohol, coal-oil, etc., are always carbon dioxide gas, and the vapor of water; and the heat developed is proportionate to the amount of these produced.

But combustion is not the only chemical source. If sulphuric acid be mixed with water, the compound becomes very hot although it is liquid. The two substances enter into an intimate chemical combination. A pint of each mixed together will not make a quart, but will fall short of that volume a good deal when they have cooled. This shows that condensation has taken place; and, knowing that condensation produces heat when brought about in other ways, one might have suspected that chemical condensation would result in a similar development of heat.

Some substances when in a finely divided state, though what we generally call solids, are capable of entering into combination with each other at a very rapid rate and then develop a great deal of heat. Such a substance as gunpowder, a combination of carbon, sulphur, and the nitrate of potash, when intimately mixed, will combine with explosive violence, and great heat results from it, as shown by the attending flash and the scorching effects it produces upon some bodies that do not happen to be destroyed by the explosion. All chemical reactions whatever involve in some degree temperature changes; and by so much one might be led to suspect that there might not be so great a difference between the mechanical sources of heat at first considered and the more obscure chemical ones as one might think who attends only to the more prominent features of the two. If one should adopt for a basis of his philosophy that like causes produce like effects, what shall he say when he sees the same effect produced by pounding with a hammer, condensing a gas, and burning a piece of wood? Either unlike causes can produce similar effects, or fundamentally these three processes are the same. We will attend to that question more at length farther on.

III. ELECTRICAL ORIGIN.

As a chapter is to be given to electricity and its phenomena, it will be sufficient here to point out that wherever a current of electricity is flowing in a conductor, there heat is invariably produced. The heat in an electric arc is so great that all known substances are either fused or volatilized in it. Gold, platinum, the ruby, are easily reduced to the liquid form, and the diamond slowly wastes away, being oxidized like a piece of coal. Electric furnaces are now in use where the most refractory substances, like clay, are reduced, and the metal aluminum extracted from it. So long as it cost so much to produce electricity as it did before the dynamo was perfected, no one could afford to use it for heating purposes. Now there will shortly be electric heaters in houses, replacing stoves for cooking and furnaces for warmth. The electrical current can be brought on the wire where it is wanted, and the heat developed from it to any degree desired. Electricity, then, is another source of heat.

IV. RADIATIVE ORIGIN.

When one stands near a blazing fire the warmth felt does not come from the heated air between the fire and the person; for when one shields his face or hands the warmth ceases to be felt, though the temperature of the air might be the same in both cases.

In like manner sunshine warms the earth, although between the sun and the earth there is an enormous space without air or other matter, through which the sun's rays come producing warmth *when they get here.* This process of giving out rays to the ether independent of matter, which is possessed by hot bodies, is called radiation. It has been shown that all bodies are at all times giving out such radiations; and oftentimes the radiation itself is called radiant energy, sometimes it is called light, and sometimes simply ether waves. Here we do not attend to the origin of the waves, but to the fact that when such waves fall upon matter they result in heating it, and therefore radiation must be looked upon as a fourth source of heat.

I would again suggest the thought presented a page or two back, as to the similarity or dissimilarity of each of these four kinds of origins of heat,—mechanical, chemical, electrical, radiant. They appear to be utterly unlike each other, yet their effects upon matter are identical, always thus and never different, so far as our experience goes. Evidently there must be some factor common to them all; and if this could be known for any one of them, it would throw light upon all the rest. If we take, for instance, the mechanical origin of heat, say, impact, which is one of the most obvious, and note the factors present, it is plain there are but two; namely, a mass of matter with a certain measurable amount of motion of the translational variety. These two embody the energy represented by the impact, and of these the translational motion is destroyed when the heat appears. The other factor, the mass of matter, remains constant. The motion that was seen needs to be accounted for; and as the heat that appears is the result of that motion, it appears probable that in some way the translational motion has been transformed into some other kind of motion, not that it has been annihilated.

TEMPERATURE.

If a pint of boiling-hot water be mixed with a pint of ice-cold water, the mixture will have all the heat there was in the pint of hot water, but it would not injure the hand thrust into it. The heat that was in one pint has been distributed through two pints, and hence each pint has one-half the heat that was in the hot pint. A red-hot bar of iron will be cooled by being thrust into a pail of water. The water will be heated, and will have all the heat the bar lost; but as it is distributed through so great a volume of water, the amount of heat in a cubic inch of it will be but a small proportion of the whole.

The word "temperature" is used to denote the degree of heat there may be in a unit volume of a substance, and this is measured by means of thermometers in which the property that heat possesses of expanding the volume of bodies is made to indicate their degree of heat. The standard for this is an arbitrary one altogether. In the common Fahrenheit thermometer there is a tube of glass with a bulb upon it filled with mercury. This, when put into ice-water, acquires the same temperature, and the mercury stands at a certain height in the tube, which is marked. Then it is put into boiling-hot water, where the mercury expands and reaches another height in the tube, which is also marked. The space between the two marks is divided into one hundred and eighty equal parts, and the same scale of division is carried beyond in both directions. A point thirty-two of these divisions below the mark of the melting ice is called zero; so between it and the boiling-point are two hundred and twelve divisions, called degrees. The centigrade thermometer is more generally used in scientific work. In this the space between the freezing and boiling points is divided into one hundred equal parts, called also degrees. A centigrade degree is $\frac{9}{5}$ larger than a Fahrenheit degree. The scales of either may be extended indefinitely for the measurement of temperatures departing from the more usual ones. For a lower limit one cannot use the mercury below about forty degrees below zero; for it freezes at that temperature, and no longer follows the same law of contraction. As alcohol does not freeze, thermometer tubes filled with it are used to indicate such low temperature. In the Arctic regions, and even in Siberia, the temperature falls to fifty or sixty degrees below zero not infrequently in winter, but temperatures have artificially been produced as low as 400° below zero.

For the higher limits mercury thermometers can be used for higher temperatures than alcohol, for the latter boils and becomes vapor at 174°. The following table of temperatures may be interesting:—

Absolute zero	−460°
Lowest degree artificially produced	−400°
Lowest degree measured in Siberia	−72°
Mercury freezes	−39°
Water freezes	32°
Blood in man	98.6°
Temperature observed in India	140°
Alcohol boils	174°
Water boils	212°
Lead melts	612°
Mercury boils	650°
Red heat visible in dark	1000°
Silver melts	1873°
Gold melts	2200°
Iron melts	2700°
Platinum melts	3600°

Gases, like liquids and solids, are increased in volume by heat when permitted to expand. If not permitted, the pressure upon the walls of the containing vessel is increased; and it is found that this pressure is proportionate to the temperature, and also that the pressure diminishes about $\frac{1}{273}$ for each centigrade degree of cooling, starting at the freezing-point of water. If, therefore, a gas could be cooled from that point 273° centigrade, it would have no pressure, as it would have no temperature. Such a degree has never yet been reached; but all phenomena having any bearing upon the subject indicate that at –273° there is no heat: it is an absolute zero. The molecules would have no translational motion, otherwise they would produce some pressure upon the walls of the vessel that contained them. Air thermometers may be made with bulbs blown upon the end of a glass tube. A small drop of water in the tube will be pushed in or out as the temperature varies, and is much more sensitive than ordinary thermometers; but barometric pressure affects it and renders it unfit for common use, but its indications are proportionate to the absolute scale; that is, the volume of the air at the melting-point of water will be increased or diminished $\frac{1}{273}$ by every change of one degree in cooling or heating, or *dfrac*1490 if the degree be Fahrenheit.

DIAG. 6.

MECHANICAL EQUIVALENT.

For a long time it was supposed that heat was a kind of substance that ordinary matter could absorb and emit. It was sometimes called caloric; and that word is in common use to-day, but not in the sense it originally had. Sometimes it was spoken of as one of the imponderables—a substance without weight. Now there is only one imponderable recognized, that is the ether. Sir Humphry Davy and Count Rumford found they could produce an indefinite amount of heat by the friction of one body upon another; and that implied if heat was a substance of any sort, that any piece of matter contained an infinite amount of heat, else one could get out of a body what was not in it. These two men concluded that heat was a kind of molecular motion, and that what their experiments showed was that friction only transformed the mechanical motion into molecular motion, which was called heat.

The old conceptions had got so thoroughly incorporated into both the thoughts and the writings of others, that they could not easily be dislodged, and men went on as they had done. It was easier to do that than to change notions and terms that were familiar for others that were strange, even if true. A whole generation of men had to be buried before any attention was paid to what had been proved in the early part of the century. Soon after 1840 it occurred to a number of persons in different countries that if heat were but transformed mechanical motion there should be some quantitative relationship between them that might be discovered; that is, a given amount of mechanical motion ought to produce a definite amount of heat, and *vice versa*. This was worked out in the most complete and satisfactory way by Joule of England. His method consisted in churning a definite amount of water and observing the rise in temperature in it. The churn paddle was driven by a known weight falling a known distance, and therefore the work done in driving the paddles was known in foot-pounds. In this way he found that 772 pounds falling one foot would heat a pound of water one degree, and he called this number the mechanical equivalent of heat. In like manner it is said that when a pound of water loses one degree in temperature, it has lost energy enough to raise 772 pounds one foot high. This relationship renders it easy to determine the amount of work a given amount of heat can do, and also the temperature that will be acquired by a given amount of water when a definite amount of work is done upon it. But the scientific importance of this new step is much greater than its practical utility. Before that time men had thought there were such things as *forces*, independent of each other; and such an idea as mutual convertibility had not dawned upon any philosophic mind. Physical philosophers were so much misled by their terminology and the accompanying notions, that Joule's work, though demonstrative, made no impression upon them for several years, and it was refused a place in the transactions of their society for seven years. The reason for this common hostility to new knowledge is probably not far to seek. When one has achieved distinction in his line of work, especially in physical science, he is likely to possess his own philosophy of things, in which not a small part of the data is symbolic and is represented in mind only by a name; and if this chances to suggest something mysterious, as, for instance, an imponderable, the less is one likely to attempt, or suffer others to attempt, to displace it by definite mechanical conceptions. To change one's fundamental conceptions necessitates a change in his philosophy throughout,—a change that is not only difficult, but highly distasteful; and one ought not to expect a welcome to a man whose work necessitates such a change.

Within the present century the advance in all directions has been such as to give definite mechanical conceptions and relations where before only ghosts and genii were supposed to do duty; and what can a man do when his genii have been slain and he must now depend upon mv^2? To become acquainted with his new associate is generally the last thing he sets himself about. It was with Joule as it was with all the prophets and discoverers. Joule, however, was young, and he lived to attend the funeral of all his detractors.

That heat and work are mutually convertible is now called the first law of thermo-dynamics; and it has led directly to a knowledge of the working-power there is in fuels, and made the duty of steam-engines and other sources of power beautifully simple.

The amount of heat needed to raise the temperature of a pound of water one degree Fahrenheit is called a *heat unit*. The amount of heat needed to raise the temperature of a kilogram of water one centigrade degree is sometimes called a calorie, and this is a unit in common use. It is found by careful experiment that a pound of coal when burnt gives up 14500 *heat units*, or would raise the temperature of 100 pounds of water 145°, or to any other equivalent. A pound of hydrogen, in like manner, burning with oxygen, will give 61000 units, a pound of wood about 7000, and so on. Each different substance has its own equivalent of such heat units. As each unit will do 772 foot-pounds of work, a pound of coal, when burnt, will give $14500 \times 772 = 11,194000$ foot-pounds of work, and so on for any other. This equivalency is independent of time or place. Whether the coal burns fast or slow makes no difference. When wood is burned in the fire it develops its work-power fast; but when it slowly rots it is undergoing the same process, oxidation, and the same amount of heat is developed, though at no time does the temperature appear to be above that of surrounding things. The food we eat possesses its mechanical equivalent, which is the maximum amount of work it would enable one to do. If bread and butter were used for the fuel of an engine, it would develop about 21000 heat units (or calories) per pound, and this is equal to $772 \times 21000 = 16,212000$ foot-pounds, and it has the same value when used for food; and thus one may know approximately the amount of energy he is supplied with from day to day; also, he may compare the amount of work he does, in lifting, walking, or otherwise, in a day with the food equivalent absorbed. Some of this is, of course, used to maintain the temperature of the body, the circulation of the blood, and so on—conditions that are tolerably constant.

THE STEAM-ENGINE.

The steam-engine is a machine for utilizing the heating-power of fuels, and, when complete, consists of furnace, boiler, and engine. The furnace transforms the energy of the fuel and air into heat units in the boiler, and the engine transforms this into the work of whatever sort it may be applied to.

Evidently the efficiency of such an engine must depend upon how large a proportion of the heat units it utilizes compared with the heat units supplied to it. Steam-engines permit the steam to escape into the air generally with a temperature higher than boiling water, and that means a great waste of unused heat; for the steam in the engine loses temperature proportionate to the work done by it, and, as stated before, the steam pressure is proportionate to its absolute temperature, not its temperature as indicated by common thermometers. And the absolute temperature on Fahrenheit scale will be found by adding 460 to the indicated temperature. Suppose, then, an engine-boiler delivered steam to the engine at 248° Fah. = 708 absolute, and on exit from the cylinder it was 212° Fah. = 672 absolute, then the proportionate amount of work done compared with the whole supplied would be $\frac{708 - 672}{708} = \frac{36}{708}$, or only about five per cent of the heating-power of the fuel. Higher efficiency must be looked for chiefly by using steam at higher temperature and, therefore, higher pressure, which would increase the value of the numerator.

The efficiency of engines is generally given in the amount of coal required to maintain one horse-power per hour. A horse-power for an hour is equal to $33000 \times 60 = 1,980000$ foot-pounds; and the coal required varies from about two pounds in the best engines to six or eight pounds, locomotive engines generally being less efficient. As one pound of coal when burnt has an equivalent of $11,194000$ foot-pounds of work, two pounds will give $22,398000$ foot-pounds. When that maintains a horse-power for an hour, or $1,980000$ foot-pounds, the efficiency is $\frac{1,980000}{22,398000} = 8$ per cent. This appears very low; but it is to be remembered that the coal is seldom anywhere near pure; that much heat escapes by the flues without heating the water; that much is lost by heating the engine, boiler, and the pipes, etc., that does no good, and most of that that does go through the engine escapes to the air without having done any work; and it cannot be helped, for steam condenses to water at 212°, and is no longer able to do steam service. In reality, such an efficiency is relatively high.

AS TO THE NATURE OF HEAT.

It has been pointed out that it was concluded early in the century that heat must be some kind of motion, because its production depended solely upon antecedent motion, and that later the quantitative relationship between the two was accurately defined. The *nature* of heat was ascertained, but the particular kind of motion that gave it its characteristics was not made out; that is, whether the motion was one of free path of the molecules,—a swinging to and fro in space,—or a true vibratory motion, such as a change of form of the molecules and atoms that made up the heated body, or a rotation of them, or a combination of any or all of these, was unknown. At first the conjecture prevailed that it was an oscillatory motion of the molecules among themselves even in a solid body; but after the discovery of spectrum analysis it became apparent that the atoms and molecules were in a state of true vibration, and their temperature depended upon the amplitude of that vibration. If one will remember that the atoms of matter are certainly elastic, and are not solid, and will also picture to himself what mechanically must happen when such a body is struck in any manner, that it *must* vibrate, for the same reason that any visible elastic body must vibrate if struck, he will see quite clearly the condition of things among elastic atoms that collide with each other so many times per second.

That they do thus vibrate is proved by the spectrum

of substances in the gaseous state where between impacts they have time to vibrate a great number of times per second. At ordinary temperatures and density a gaseous molecule of hydrogen, having a mean free path of about the two-hundred-and-fifty-thousandth of an inch, and moving at the rate of 6000 feet per second, will collide with its neighbors 17750 millions of times per second, but its spectrum shows that it makes 450 millions of millions of vibrations in the same interval, so that in each interval between impacts it would be able to make $\frac{450,000000,000000}{17750,000000} = 25352$, more than twenty-five thousand vibrations.

Now, imagine a number of bells suspended by cords of equal length from the ceiling, but not so near as to touch each other. Suppose each bell to have the same musical pitch as every other one, and now let one of the outer ones be pulled away from the rest and forcibly swung back among them; presently every bell among them would be set swinging by the impact of others upon it, and each impact would cause each bell to sound its own particular pitch, and the elasticity of each individual one would maintain that vibration in some degree until the next impact, when it would be strengthened, and one would hear along with the bumping of the bells the sound due to the pitch of the individual bells. Something very like this goes on among the molecules of the gas. Their vibratory movements we cannot hear, but with the spectroscope they are detected and measured. Now, hot bodies cool by radiation—the giving-off of just such waves in the ether as we are describing,—and the fact that such cooling molecules of a gas give out constant wave-lengths, as is shown by their spectrum lines, is proof that the vibrations that originate the waves are not free-path or oscillatory motions, but true atomic ones, due to a *change in form*. How this can be is easily seen by considering the change in form made by any vibrating body, say, a ring. Let the heavy lined ring represent an elastic atom: if it be subjected to impact it will assume an elliptical outline, and go through a series of phases represented by the dotted lines. This change of form, and uniform vibration, is a mechanical necessity, and is independent of the size or particular form a body may have. It is this kind of motion that embodies the energy represented by the temperature of an atom or a molecule, and the temperature varies with the square of the amplitude of this motion; and two bodies have the same temperature

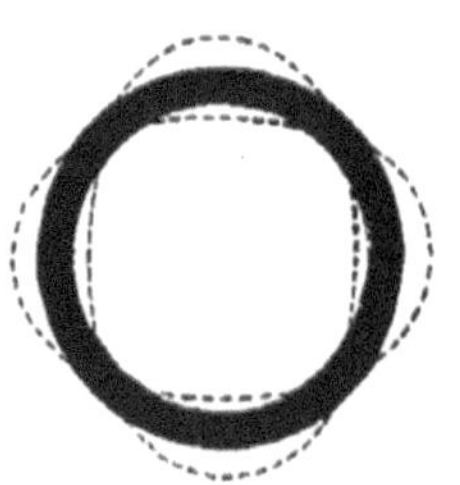

DIAG. 7.

when their molecules have the same vibratory energy. A single molecule in free space would radiate all its heat away, and thus be reduced to absolute zero, if it were not continually receiving from other bodies an amount that depended upon its nearness to them and their own amplitude of similar motion. Hence the temperature of a body depends upon the amplitude of vibration of its molecules, and not upon any translatory or oscillatory or rotatory motions. This is not saying that molecules that are heated do not have other motions than the vibratory ones constituting their temperature, but when they do have others it is at the expense of the vibratory, and therefore has reduced the temperature; and such free-path motion as all gases have, and which produces pressure upon the walls of vessels, is maintained by the vibratory. It is not heat, but the result of heat, in the same way as the translatory motion of a bullet is not heat, but the result of heat. Most books on heat do not make the distinction here made, but combine the heat-motion of the molecules themselves with the translatory motion they have, calling the sum of them the heat of the gas. So long as one is concerned only with the energy involved in the actions it will make no difference; but if one analyzes the process for the factors it is plain that there are two distinct kinds of motion—one of them capable of setting up waves in the ether, the other not, for it is not known that any free-path or translatory movement of a body ever disturbs the ether; and if distinctions of such marked characters as these exist, and one of them involves temperature and ether waves, and the other does not, they ought not both to be called by the same name. The peculiar character of the energy involved in heat as distinguished from so-called mechanical energy, is that the factor of motion is of the vibratory sort, whereas the other is more or less translatory,—one capable of easy transformation into ether waves, the other incapable of such transformation, but each of them easily transformed into the other by impact. Equivalent velocities give the same amount of working ability, or $\frac{Wv^2}{2g} = \frac{Wa^2n^2}{2g} = Pd$ (see p. 82). So it can be understood how ordinary visible motion can be transformed into heat, and *vice versa*, as easily as one can understand how the motion of the clapper of a bell is transformed into sound.

ORIGIN OF THE SUN'S HEAT.

There has been much speculation as to the source of the heat of the sun. Unless one assumes that it has some miraculous or non-physical origin he is bound to account for it, if at all, upon the assumption that physical conditions and relations, such as we find at the earth, hold good at the sun as elsewhere.

At the beginning of this chapter the various sources of heat were considered,—the mechanical, the chemical, the electric, and the radiative. If these be tested as to their sufficiency to account for the temperature of the sun, one may reach a conclusion as to the probability of any or all of them being concerned in it and their relative importance.

It will be convenient to consider them in the reverse order, and first as to radiations. In order that a body should become heated by radiations, there must first be some body or bodies having as high or higher temperature to give rise to the radiations; and in this case, if the sun's heat came from such a source, one would need to look for the other bodies in the universe having such high temperature. The millions of stars shining by their own light would at first seem to furnish the proper source; for the testimony of the spectroscope is that they all are highly heated, and some astronomers think some of them to be much hotter than the sun is. One of the conditions under which radiant energy is distributed in space is that its amount upon a given surface is inversely as the square of the distance from the source; and as every one of these bodies is at such an amazing distance away, it is only with the most delicate instruments that their radiant energy can be measured, and a given surface upon the earth would receive as much as the same surface upon the sun, and the earth would be heated from the same source as much as the sun would be. Practically it is found to be but a very small quantity, and hence radiation from other bodies cannot possibly account for the sun's heat.

Second, as to electrical currents: it may be said at the outset we have no direct knowledge that there are such at the sun, and from other knowledge we have of its constitution it would appear to be highly improbable that there were or could be electric currents there. Electric currents imply some generator and some conductor for their transference; and from what is known or may fairly be inferred that every substance we are acquainted with as a conductor of electricity which is present in the sun—and there are a good many of them—iron being particularly abundant, yet they are all at such a high temperature as to be a far reach from the conductibility we know anything about. There may be, but it is by no means certain, something solid in the sun, but the most of it is as gaseous as a bubble, and gases do not conduct currents of electricity.

Third, chemical action is known to be the antecedent

of vast quantities of heat. It may be recalled that a pound of hydrogen, for instance, when allowed to combine chemically with oxygen will give out 61000 heat units. The atmosphere of the sun appears to be made up of elements mostly in an uncombined form, except in the cooler, outlying parts; that is, the temperature is so high that chemical combination is impossible except in exposed places where radiation can allow cooling to take place. It is tolerably certain that chemical combinations are taking place there whenever it is possible, and with such combination heat must be produced, if physical laws are in operation there as they are at the earth, but the amount of it going on, or possible, if the whole body of the sun were to combine its elements in this way, does not appear to begin to be equal to the expenditure of heat actually taking place.

There remains only the mechanical sources of impact, friction, and condensation. There is good evidence that there is a large body of meteors in the neighborhood of the sun that must be falling upon its surface. The sun's attraction can give a velocity of nearly four hundred miles a second to any body reaching him from distant space, and such a velocity would, on impact, produce heat enough to reduce the whole body to a gaseous state almost instantly.

Given the mass and velocity of a body, and one may calculate how much energy it has, and how much heat is the equivalent of the mechanical energy. Such a computation shows that even if the earth were to fall into the sun, it would be volatilized in a very brief time. If the sun's surface were solid the impact would be sufficient to effect it almost instantly. If the shell of the sun were liquid it would be changed more slowly through friction, but, in the end, the result would be the same. It does not appear, however, that there is sufficient material that finds its way to the sun to furnish but a small proportion of the sun's heat, so neither impact nor friction can be admitted as sufficient agencies. There remains but one more, namely, compression. Is there any evidence that condensation is taking place? The body of the sun is 866000 miles in diameter, and is so far away that this immense magnitude occupies but about half a degree of arc. If it were to shrink at the rate of a mile in twenty years, it would account for the present rate of expenditure, but such a shrinkage could not be observed from the earth for several thousand years, for nothing much less than a second of arc can be observed with certainty, and a second of arc at the sun's distance is equal to about 465 miles, so it would require $465 \times 20 = 9300$ years to produce an observable effect.

Now, if the nebula theory be true, the sun once occupied all the space between itself and the outer boundary of the solar system and has shrunk to its present dimensions, a process which, if heat alone were concerned, would require about eighteen millions of years. It is not probable that heat alone has been concerned, so it is probable that the sun is older than that, but the shrinkage will account for the heat, and it appears as the only probable conjecture. It will be understood that the gravitative action is the occasion of the compression, and that the approach is constant and as fast as the generated heat can be radiated away. It has been calculated that at the above rate of condensation it may be reduced to one-half its present diameter with its present radiation rate, in about five million years, when its density will be about twice that of water.

From such considerations it appears in a high degree probable that the heat of the sun is due to condensation, the condensation is due to gravitation, and thus one is led back to a time when the substance of the sun and all the planets was scattered through that immense space, the diameter of which is not less than six thousand millions of miles. How matter came to be thus scattered is at present an enigma. It is important to remark here, though, that until there was impact among atoms, and molecules were formed, there evidently could be no such condition as what we call heat, and until these atoms and molecules vibrated there could be no light, that is, ether waves.

EFFECTS OF HEAT.

Once in possession of a good, mechanical conception of the action going on in a heated body, one can proceed to trace out the various effects of heat in all directions. Thus to take the familiar one of pressure in a gas. A gas is simply a large number of individual molecules moving about with great velocity and bumping against each other and the sides of the containing vessel. Each molecule, though small, has some momentum; but the enormous number of them in, say, a cubic inch, five hundred millions of millions of millions, and their relatively high translatory velocities,—say fifteen hundred feet per second, gives them momentum which, when spent upon the side of the vessel, gives a pressure equal to about fifteen pounds per square inch. If one were to hold up a shield against which many balls were thrown per second, he would need to brace himself to withstand the pressure that would appear to be constant.

If the gas be heated the molecules have increased

amplitude of vibration, and they rebound from each other with greater velocity, and strike the side with more momentum, and hence the pressure is greater. As the pressure is proportional to the absolute temperature, it is plain there could be no pressure if there was no vibratory motion. If the density of the gas be increased by adding more molecules per cubic inch, there must a greater number of them strike upon the sides of the vessel in a second, which will increase the pressure, that is, the pressure varies as the density.

When it is said that gases have a tendency to expand, or that they exhibit a repulsive action, all that is signified is this; as elastic bodies, the molecules rebound after impact, and continue on in their direction, according to the first law of motion, until otherwise obstructed. When a ball rebounds from the side of the house it has been thrown against, it is not because there is any repulsion between the ball and the house.

EFFECT OF PRESSURE UPON BOILING AND FUSION.

When it is said that the boiling-point of water is 212°, it is to be understood that the pressure of the air upon the surface of the water is fifteen pounds per square inch. At elevated places water boils at a much lower temperature; and when in a tight vessel, like the boilers of steam-engines, the pressure of the steam affects its boiling-point in the opposite way, raising it. Thus at twenty pounds steam pressure, the temperature required to boil water is 228°, at sixty pounds it is 291°, at ninety pounds 319°, and at the high pressures employed in locomotives of one hundred and fifty pounds or more to the square inch, the temperature of the steam and water is 360° or more. As one goes down into a mine the pressure of the air becomes greater, and higher temperature is needed to boil water. The explanation of this phenomenon is that the heated molecules of the liquid are bumping against each other in all directions, but the surface molecules can receive such bumps only from below and on their sides. If there were no molecules above to beat downwards, the surface molecules would fly rapidly up into the free space, which would be what we call a vacuum. This escape of the surface molecules of a liquid into the space above is called evaporation, and the higher the temperature of the liquid the harder the bumps, and the more will be flipped away from the liquid and become free rovers, having a long, free path. When, however, the gaseous particles are numerous and strike back upon the surface, that is, when there is a gaseous pressure upon the surface, the surface molecules are prevented from rising, that is to say, evaporation cannot go on so fast, boiling is prevented until more energy is given to the water, and that means a higher temperature.

The melting-point of substances is likewise affected by the pressure to which they are subject, and increasing the pressure increases the temperature needed to fuse them. Such small variations of pressure as only a few pounds per square inch do not make much difference, but pressure measured by tons per square inch makes a great deal. The condition of the interior of the earth appears to depend upon this as a most important factor. As one goes beneath the surface of the earth in mines and tunnels, it is observed that the temperature rises about one degree for every fifty or sixty feet of descent; and it was formerly inferred from this that at the depth of a few miles a temperature would be reached high enough to melt the most refractory bodies, and hence the interior of the earth was probably in a fused state while the crust was relatively thin. Such a view took no account of the effect of pressure upon the state of bodies. At the depth of a mile of water the pressure must be equal to $62.5 \times 5280 = 330000$ pounds per square foot, and as rock is $2\frac{1}{2}$ times the weight of water, the pressure must be 825000 pounds, or over four hundred tons; and at five, ten, or a hundred miles, it is obvious the pressure is correspondingly greater. A body that at the surface of the earth would melt at any assignable temperature would require a much higher temperature to fuse when subjected to such enormous pressure. It appears that the pressure increases faster than the observed temperature; and hence the earth must be solid to the centre, instead of being liquid as formerly supposed. This makes it appear that the phenomena of volcanoes are only local, and do not indicate any general melted condition of the earth. If a body that would melt at a thousand degrees on the surface of the earth be subject to such pressure that it is not melted when its temperature is two thousand degrees, then, if the pressure be suddenly removed from it, the heat it has will instantly liquefy it. This may be the condition at the base of volcanoes, where shrinkage of the earth's crust in some direction may relieve the pressure in some other direction; and a large mass of heated material may become liquid, expanding in volume, and overflow in any direction where there is a vent, and this would be called a volcanic eruption.

MAXIMUM TEMPERATURE.

We have considered the condition called absolute zero, wherein the molecules have no vibratory motion whatever; and it has also been pointed out, and it is generally agreed, that the temperature of a body varies as the square of its amplitude of molecular vibration.

It has often been assumed in treating of high temperatures, such as that of the sun for instance, that there is no limit to the temperature to which matter can be raised. So some have estimated the temperature of the sun to be several millions of degrees; but a consideration of the factors involved will show such a conclusion to be impossible, for the dimensions and form of a body set a limit to the amplitude it can have. A tuning-fork cannot have its prongs vibrate beyond the limit where its prongs touch each other, and a vibrating ring cannot have an amplitude greater than one-fourth its circumference; and this degree is only possible to a mathematical circle having no thickness. Make a ring of a piece of twine, and elongate any diameter until the opposite sides touch, then move the middle points through a similar distance, and it will be seen that the limit will be equal to a quadrant of the circle; but if the ring be a thick one, say made of rope, it would be less than that, and how much less will depend upon the relative thickness of the rope to the diameter of the ring. If the thickness of the rope were one-fourth the diameter of the ring, then the amplitude could be but one-half the quadrant, and so on. Now, the atoms of matter have a definite size, and no one has ventured to suggest that they were variable in size in any degree; and one may, therefore, compute the greatest amplitude such a body could have, whether it were a circle or a hollow sphere without thickness. If the diameter be as before stated, one fifty-millionth of an inch, calculation shows that the greatest amplitude it could have would be about one sixty-four-millionth of an inch. This, multiplied by the number of vibrations it makes per second, will give the equivalent velocity from which its energy can be calculated. On page 79, it is shown that the velocity of a vibrating atom, if the amplitude be one-half of the diameter, will be about eighty miles a second. If the amplitude be equal in measure to the quadrant, as is here supposed, this velocity would be not far from a hundred miles per second, and the energy represented by that velocity would be the utmost energy of heat, or highest temperature that the body could have. The pressure of gases enables one to determine the velocity of the particles; and when this is known at a given temperature, the temperature at any other velocity may be computed.

The statement that atoms and molecules can have a maximum temperature must not be understood to imply that the energy they can have is fixed at that limit, because aside from their temperature energy, represented by their vibratory motion, they can have any assignable translatory velocity in addition. But it does imply that ether waves, arising from temperature, have a fixed limit for each element; and such radiant energy from a given source cannot be transmitted beyond a certain rate, because its amplitude has a limit, so that whatever actual energy the sun as a whole may have, it cannot lose that energy by radiation faster than an assignable rate.

This has an important bearing upon the question of the age of the sun. Computations have been made of the length of time the sun can have been giving out its energy, on the assumption that the sun is a cooling body, and that it was formerly much hotter than it is now. If the above statements are correct, the probability is that the sun is as hot now as it ever was, and that its rate of loss of heat by radiation has not been greatly different from what it is to-day; so, instead of being only fifteen or twenty millions of years old, it may be very much more.

As the temperature of a body represents its molecular energy, and is measured by $\frac{mv^2}{2}$, it follows that if two different kinds of molecules, such as hydrogen and oxygen, have the same temperature, they will have the same amount of energy; but the mass of an oxygen molecule is sixteen times greater than the mass of a hydrogen molecule. In an equal weight of the two there will be sixteen times more molecules of hydrogen than of oxygen, and therefore the hydrogen will have sixteen times the energy of the oxygen at the same temperature. To produce a rise of temperature of one degree in a pound, or any given weight of hydrogen, would require sixteen times as much heat as the same weight of oxygen would need. This difference in thermal capacity of different substances is called their specific heat. In general, the lighter the molecules that make up a substance, the more numerous must they be to make up a given mass, and the higher will be its specific heat; i.e., the more heat must be expended upon it to produce a given rise in its temperature. The specific heat of water is chosen as a standard and is unity, as it is found to require more heat to raise a given weight of it one degree than any other substance. One heat unit will raise the temperature of a pound of it one degree; all other substances require but a fraction of this. From what is said, it appears that the specific heat of an element varies inversely as its atomic weight. The specific heat of a substance determines the temperature it will attain when a definite quantity of heat is supplied to it. If a pound of hydrogen and eight pounds of oxygen are exploded together, and not allowed to expand in volume, 51444 heat units calories are produced. The 51444 heat units would be divided among nine pounds of water vapor, that has a specific heat under such conditions of .37. The temperature attained would be $\frac{51444}{9 \times .37} = 15450^\circ$. This temperature is much higher than the limit of possible combination of the two gases, which, at about 3000°, are unable to combine, so such an action could not take place any faster than the parts could cool down to the latter temperature. If the mixture be allowed to expand, the temperature of 3000° may not be reached, and the action of the whole is so rapid it is called an explosion.

DISSOCIATION.

When compound molecules are broken up into their elementary constituents in any manner, the process is called dissociation. It may be effected by electrical action, as when water is decomposed by it, or by chemical action, as when wood is decomposed under water, setting the carbon free; but heat is competent to effect the same end. At the temperature of about 3000° the existence of water is impossible, as the elements cannot stay united, and the reason is obvious. Whatever the nature of the attraction that holds atoms together in chemical compounds, if the elementary atoms are themselves in brisk vibratory motion, as we know they are, they must be straining their bonds continually to separate; and when the amplitude of such motion reaches a certain maximum, the impacts are so violent as to make the atoms rebound out of each other's neighborhood, and thus prevent cohesion. The atoms then either enter into new combinations with others, if possible, and if not they remain as gaseous particles, and subject to the laws of gases.

If one starts with a piece of ice and applies heat it melts, and we call the liquid water. Apply more heat and the water becomes steam, in which the individual molecules are no longer able to cohere, because of their energetic motions; but each molecule remains intact, having a long free path, for a cubic inch of water becomes nearly a cubic foot of steam under ordinary air pressure. If still more heat be applied, the molecules become more and more unstable until they too are broken up in the same way and for the same reason that the solid and the liquid forms were. When it is no longer possible for hydrogen and oxygen to combine, it is still possible for the atoms of each to combine with each other, hydrogen with hydrogen and oxygen with oxygen, forming elementary molecules *H H*, and *O O*; but if a still higher temperature be applied, even this combination becomes impossible, and the atoms themselves become free rovers and individually independent. Thus it is seen that the different states of matter depend altogether upon temperature. At absolute zero there can be no such thing as a gas, for the molecules would have no individual vibrations and therefore no free paths. They would probably fall to the bottom of the vessel and remain quiescent. It is also probable that both liquids and solids too would cease to exist, not that matter would be annihilated, but a solid, a liquid, and a gas are simply each a bundle of physical properties that depend mostly upon temperature, and those properties would probably disappear with the disappearance of the conditions upon which they depended.

CHAPTER VII

Ether Waves

It has already been stated in what has preceded this that translational motions of matter are not competent to originate ether waves, and that vibratory motions of both atoms and molecules can originate them. A consideration of the origin, transmission, and effects of such ether waves constitutes the subject-matter of what is called the science of light. The word "light" is commonly used to signify that agency in nature which is capable of affecting the eye and causes vision, or the sensation of sight, and until within a very few years has been supposed to be a peculiar kind of a wave motion in the ether quite distinct from other waves known to exist which were competent to produce heating and chemical effects, so such waves as were known from their effects were called heat waves, light waves, and actinic or chemical waves, according as they heated bodies, produced light, or brought about chemical reactions. These three sorts of waves were supposed to coexist generally, but were capable of being separated from each other so there could be a beam of either without the others. This is now known to be a mistaken view, for what a given ether wave will do depends upon what it falls on rather than on its own peculiarity. The same waves that fall upon the eye and produce the sensation of sight will heat other kinds of matter, and if they fall upon a surface of molecules that are unstable, that is, in which the atoms that make up the molecules are not strongly cohesive, the molecules are disrupted by the waves, and the atoms enter into new combinations, and this process is called a chemical process; and while it is true that some waves will not produce vision, there are none that will not produce both heating and chemical effects, so there is no such distinction among ether waves as was supposed, and this leads to another conclusion also; viz., if there is no such distinction between waves, then there is no such thing as light at all, unless we classify all rays as light, whether they can produce sight or not, which is sometimes done to save explanations, but it leads to the anomaly that there is such a thing as dark light, which is absurd. There will be no difficulty whatever if light be defined as a sensation merely, and the waves competent to produce the sensation be called visual waves. Up to the present time, however, the old terminology is quite generally adhered to in spite of the difficulty of reconciling the old signification with the new knowledge. There is no single word that signifies ether waves in general, and independent of the effects that may be produced in specific cases, and for that reason this term has been adopted. The word "light" is entirely inadequate, and likely to mislead one not well versed in the phenomena.

ORIGIN OF ETHER WAVES.

The source of ether waves of all degrees whatever is the vibratory motions of atoms and molecules as distinguished from their translatory, or free-path motions, but their rates of vibration are determined by their atomic weights. An atom of hydrogen, for instance, has a different rate from oxygen, for the same reason that two tuning-forks, though made precisely alike, would have different rates if one were made of steel and the other of aluminium. If they have different rates, then the number of waves produced by them per second will be different, and as all waves travel in the ether with the same speed, namely, 186000 miles per second, the length of the waves produced by them must be different.

There are about seventy different elementary atoms, each setting up its characteristic waves in the ether all the time. It is to be remembered that all atoms and molecules are always to be considered as hot bodies; that is, bodies having some temperature, and mostly a long way above absolute zero; and also that their energy of this kind may be spent upon the ether. If the waves from one molecule have more energy than those given off by a second molecule upon which they fall, the second one absorbs some of it so as to have its own temperature raised until it is the same as the other; that is, until the energy given off by them both is equal. And this is universally true. Matter is continually exchanging energy in this way, always tending to bring about equality of temperature. But the number of vibrations a body makes does not need to be the same as that of another body in order to possess the same amount of energy, for the energy depends upon both mass and velocity. If the mass be small, the velocity must be greater, and *vice versa*. And thus it is that the seventy elements that make up the kinds of matter we know are everywhere and at all times setting up ether waves, each kind its particular rates, when not otherwise interfered with.

There is, however, a qualification that must be added that has a high degree of scientific importance. Every elementary substance is vibrating at several rates at the same time, as do piano-strings, bells, and musical instruments in general. Every particular rate of vibration produces its own waves, and thus each atom and molecule is continually producing, when not interfered with, its own characteristic set of waves. This must make the ether waves from the different kinds of matter exceedingly complex, and disentangling them correspondingly difficult; yet it has been done.

When we look at luminous bodies, like the sun or stars, or flames, or gas, they seem to differ from each other in brightness and sometimes in color, as is seen in fireworks. A flame of alcohol has a bluish tint, a little salt in it makes it yellow, some lithium makes it red, and copper, green or bluish, while sunlight is white, as is the electric light. If one looks through a common prism at the landscape the edges of objects appear in rainbow tints, and with the colors arranged in the same order, while at the same time the shape of things is more or less distorted. If a beam of sunlight be sent directly through such a prism, a patch of colors may be seen on the floor or wall, and this is called a solar spectrum; and if this light of different tints has its wave length measured, it appears that the red light has a wave length of about the one forty-thousandth of an inch, and the violet light at the other extreme a wave length of about the one sixty-thousandth of an inch, while the intermediate tints range regularly from the one to the other. There is in this spectrum that can be seen an almost infinite number of wave lengths; there is no break among them apparently. The same

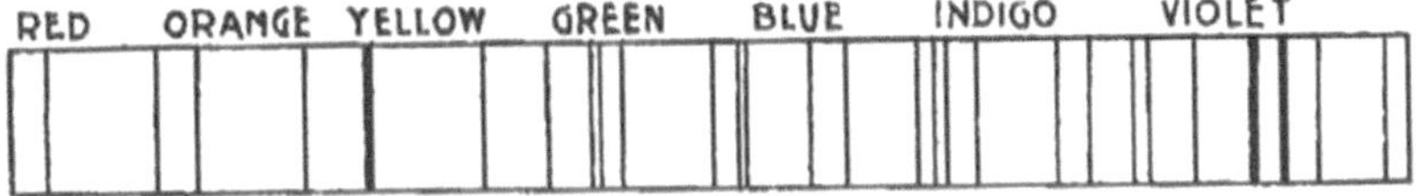

DIAG. 8.—VISIBLE SOLAR SPECTRUM.

thing holds true of a spectrum produced by letting the light from a lamp or candle go through the same prism: the tints, their order, and their wave lengths are found to be the same. The prism then receives ether waves of any or all wave lengths, and separates or disperses them in the order of their wave lengths. In doing this it deflects the longer waves less than it does the shorter ones. The deflection of the waves from their original course is called *refraction*, and the separation from each other so as to produce the spectrum is called *dispersion*. A prism effects both at the same time, and thus enables one to isolate at will any particular tint or part of the spectrum; and if one takes a single narrow portion in any such spectrum, he has a bundle of light rays of uniform wave lengths, and he may then determine their value. In this way the wave lengths of the different colored parts of the spectrum of sunlight have been found to be as follows:—

Red,	about	39000 to the inch
Orange,	"	41000 " " "
Yellow,	"	44000 " " "
Green,	"	47000 " " "
Blue,	"	51000 " " "
Indigo,	"	54000 " " "
Violet,	"	57000 " " "
Extreme visible, about		60000 " " "

A spectroscope is an instrument composed of a prism mounted between two tubes, one of them having an adjustable slot for the light to be examined to pass through on its way to the prism, the other being a short telescope to magnify somewhat the image of the spectrum that it may the better be seen. With this, light from any source may be examined. Light made up of all wave lengths that can be seen shows as a complete spectrum, while any light made up of but a part of these gives a corresponding incomplete spectrum. The flame of an alcohol lamp, or a Bunsen gas-flame, gives but little brightness and not much to produce a spectrum; but a little salt in the flame gives to it a bright yellow tint, and shows in the spectroscope a single narrow band of yellow light in the same place as the yellow seen in sunlight, and therefore having the same wave length. Such a beam made up of waves of one wave length is called homogeneous light. This sodium light has a wave length of about

DIAG. 9.—SPECTROSCOPE.

the one forty-four-thousandth of an inch. With other more refined methods, which cannot be described here, sodium is found to have other wave lengths beyond both the red and blue ends, and which cannot be detected by the eye alone. Hydrogen, another element, gives a bright red line and a blue line that are easily seen; and several others may be detected with more delicate apparatus. In this manner all the elements have been attentively studied during the past thirty years, and many treatises may be found that give full particulars of the processes and results. The substance of knowledge obtained by the study of the spectra of the elements may be briefly stated to be,—

1st, Each element has its own vibratory rates at a given temperature, and sets up corresponding ether waves; some of these can be seen, and others require more complicated apparatus to discover.

2d, In order that the characteristic vibrations of any atoms or molecules may take place, it is necessary that they be allowed a free path to vibrate in; in other words, they need be in the gaseous state. If they be crowded together, as they are in solids and liquids, they have no chance to vibrate without interference. A pailful of school-bells might make a jangling noise, but would give no particular pitch or characteristic sound of any of the bells, and only when not interfered with for a part of the time at least could one give out its true sound. This gaseous state is generally obtained by igniting in flames or by the electric spark the substance to be examined. In an electric arc all substances are volatilized, and may be then studied with the spectroscope to great advantage. Sometimes substances that remain in the gaseous state at ordinary temperatures, such as hydrogen, oxygen, chlorine, etc., are hermetically sealed in glass tubes, after rarefication, in order to obtain long free paths, and are lighted up by means of electric discharges through them.

3d, On account of the lack of vibratory freedom,

the molecules of solids and liquids give out vibrations of all wave lengths, for every partial and incompleted movement disturbs the ether; and there are all degrees of these, but the energy of the shorter ones is rarely great enough to affect the eye, and hence are not visible at ordinary temperatures. If a body like a cannon-ball be gradually heated in the dark, it will presently begin to glow with a dim red tint. If looked at through the spectroscope, only red light on the extreme red border can be seen. As the temperature rises, additional shorter waves appear, and the spectrum broadens to the orange, then the yellow, and so on; the ones already showing growing brighter meanwhile, until the ball is in a bright glow, and a full continuous spectrum is produced. As the ball cools, the reverse holds true; and the violet waves are the first to disappear, then the blue, and lastly the red vanishes from sight. Still the ball is much too hot to safely touch, and continues to cool by giving off ether waves differing from the rest only in being too long to affect the eye. They still are refracted by the prism, and an invisible spectrum is produced, and this spectrum has been traced out to ten times the length of the visible spectrum.

The sun, an electric arc, and other solid hot bodies, give out similar long, invisible spectra.

In like manner, where the body is white-hot, and

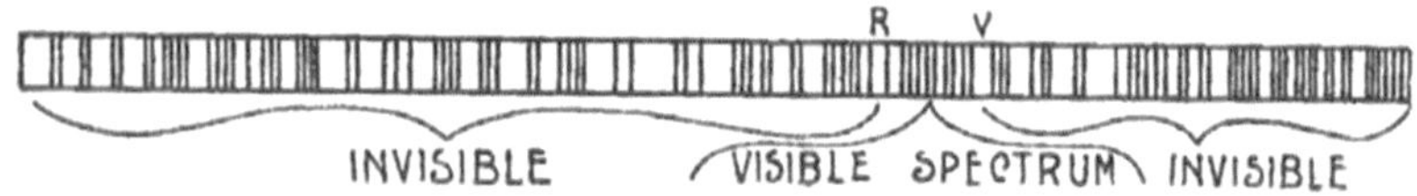

DIAG. 10.—COMPLETE SOLAR SPECTRUM.

giving out the shortest waves the eye can see, there can still be found, a long way beyond that limit, waves that can do photographic work, which is but a kind of molecular dissociation.

4th, Where waves of a given length are made to pass through a gas having similar vibratory rates, or capable of producing waves of the same length, the molecules of the latter will absorb such waves, and therefore stop their progress, especially if they have more energy than the waves the absorbing gas can give out. So if sunlight containing the same yellow light as that of sodium gas be made to pass through the latter, it will be stopped; and if this be done where there is a spectrum of sunlight, the yellow will be cut out from it, and there will be but a black line instead. This is called gaseous absorption, and is an illustration of what was said a little way back about the exchange of energy always going on. The absorbing power of a gas has a significance like its radiations, and indicates its presence as well.

The yellow light of sodium gas has a definite place in the spectrum; and hence if one perceives those wave lengths in a gaseous spectrum, he knows that sodium must be present in a state of incandescence, giving rise to the waves. But if the light from a white-hot cannon-ball were to be sent through that same vapor, and afterwards examined with a prism, the yellow light would be absent, and the absence would still proclaim the existence of sodium vapor.

Hence, if an incandescent body gives a continuous spectrum, it must be a solid or a liquid; the molecules must be so compact that the individual vibrations are prevented, and only irregular ones can be made. If a discontinuous but bright line spectrum is shown, the matter must be in a gaseous state, and the molecules have free path.

If a bright spectrum have black spaces or bands across it, there is indicated a solid or liquid incandescent body shining through gas that acts by absorption upon it, and thus both the solid and gaseous conditions are detected, as well as the nature of the substance in the gaseous state.

This knowledge has been applied to the discovery of the substance and condition of the sun and other celestial bodies, and it is concluded that the sun has a solid or liquid surface as a shell to a gaseous interior, and that the atmosphere of it consists of the various elements that make up the body of the sun in so highly heated a condition as to keep them in a vaporous or gaseous state. The characteristic spectroscopic lines of about forty elements have been found there. Some of the elements have a very large number of spectroscopic lines. Iron, for instance, has several hundred lines. Hydrogen is particularly abundant. Perhaps the most important discovery due to the spectroscope has been this: that there are a very large number of gaseous bodies, called nebulæ, in the heavens; some of these fill immense spaces; they are in a condensing state, and all of them are mostly made up of hydrogen. This discovery gave an additional probability to the nebula theory of the origin of the solar system, for it showed that process in its various stages in more distant parts of space: and in addition to that, it has led to the surmise that in some way some of those we now call elements are really compounds of more elementary substances, probably hydrogen; but that is a speculation merely, for there is no other than such spectroscopic evidence that anything like transmutation of what we call elements into others can take place.

The spectroscopic examination of the other members of the solar system has shown that Mars has an atmosphere like ours, holding watery vapor in it; that

Jupiter is red-hot; that the temperature of Saturn is probably much too high for any such living things as exist on this earth—and in this way has answered the question so interesting to most thoughtful persons as to whether the planets are inhabited or not. Jupiter certainly cannot be inhabited by any such beings as we are, for the temperature would destroy all organic things.

Velocities of translation can also be measured when as high as two miles a second or more, by the displacement of spectroscopic lines towards one or the other end of the spectrum. If a star is approaching us, the wave lengths are shortened a small quantity, and that changes the position of a line towards the blue end, while recession makes it longer and moves it towards the red end, so it has been found that Sirius is receding at the rate of nineteen miles per second; that Arcturus is coming towards us at the rate of sixty miles per second. In like manner is shown that the sun, and with him the whole solar system, is travelling in the direction of the constellation Hercules at the probable rate of about sixteen miles per second.

Now, all this presupposes that the principles established in the laboratory for substances there investigated are applicable wherever such matter exists; for instance, that the spectrum of sodium and of hydrogen and iron, which depends upon temperature and pressure, is as reliable if the light comes from a body a million miles or a thousand million miles away as if it came from only one mile or a foot distant. If it be thus widely applicable, then do we have the best of testimony that matter, its conditions, and its laws are the same everywhere, and that the earth is a fair specimen of the rest of the universe.

OTHER PHENOMENA OF ETHER WAVES.

Whenever a line of ether waves—which is generally called a ray, whatever the wave length may be—falls upon matter, the ray may be either absorbed, transmitted, or reflected. Neither of these results takes place singly in any case. There is no known body, for instance, that can wholly absorb all the rays that fall upon it, nor wholly transmit or reflect them. If a body should be able to absorb all the rays that fall upon it, we should not be able to see it unless itself were a self-luminous body, for we only see other than self-luminous objects by means of the light reflected from them, and such a body would reflect no light, and hence could not be visible.

Bodies which absorb most of the rays that fall upon them we call black and opaque; that is, a body that reflects but a small portion of the waves that are incident upon it is a dark or black body, because we see but little of it. If it reflected none at all, it would be quite invisible. In like manner, a perfectly transparent body would be one that would neither absorb nor reflect any rays, and for that reason would be quite as invisible as space itself. The air is perhaps as near an approach to perfect transparency as anything that can be named; yet if it reflected no rays at all, there would be nothing of the diffused light that is now so plentiful on the clearest day, but there would be only what would come direct to us from the sun or other luminous body. We call clear glass and water transparent because objects can be plainly seen through them; and a sheet of hard black rubber we call opaque, for nothing whatever can be seen through it, nevertheless it has been shown that waves longer than those that affect the eye, go through such hard rubber as easily as the shorter ones we call light go through glass, hence transparency and opacity are terms only relative to particular kinds of waves. All kinds of matter reflect more or less of the waves that fall upon it. This reflection is merely the change in direction of the ray; but it always follows a definite law, keeping to its original plane, and making the angle of reflection equal to the incident angle. The surfaces of most bodies are very rough, and the rays are reflected in all directions, because the points upon the surface face in so many ways. This will be obvious to one who looks at the surface of paper or of wood with a magnifying-glass. The smoother a surface is made, the nearer will all the incident rays take the same direction on reflection. Mirrors are thus made of smooth glass or metallic surfaces, and are plane, convex, or concave; but whether they are made with plane or curved surface, the rays reflected always follow the above law.

REFRACTION.

So long as ether waves fall perpendicularly upon any surface of any kind of matter, the rays go straight on into it if they be not reflected or absorbed at the surface; there is no change in the direction, but the velocity of transmission is less in all kinds of matter than it is in the ether. In glass it is only about two-thirds as fast, and in water about three-fourths. When the ray meets the surface at an angle, it is bent out of its course more or less, depending upon the kind of material it falls upon, and also the angle at which it meets it. This change of direction, when entering a new medium, is called refraction, and this property is possessed by all kinds of matter, solid as well as liquid and gas. The refraction for a given angle of incidence is more for a liquid than for a gas, more for a solid like glass than for water or other liquids, and more for a diamond than for any other known substance. The same rule that obtains when the waves enter a medium, holds when it leaves it; the direction it will now take will depend upon the angle the rays make with that surface and the character of the medium into which it enters. Thus, if a ray meets a piece of plain glass at an angle, say, of 45°, some of it will be reflected, making an angle of 90° with the incident ray, and some of it will be refracted into it, making an angle with the original direction, and continue on in a straight line until it meets the next surface, when it will again assume its original direction: but when the second surface is not parallel with the first, as is the case with the prism, the direction may depart still more from the original; and the shorter the wave length, the more the deflection. It is this property that is made use of in spectroscopes, microscopes, and telescopes. A lens has one or both surfaces curved, so as to be convex or concave, depending upon the use it is to be put to,—a convex glass converging the rays, and a concave one separating them,—and almost any degree of either of these may be obtained by proper curvature.

Both microscopes and telescopes are so common, and descriptions of them are to be found in so many places, that they need not be described here. The inquiry is often made, why still more powerful microscopes and telescopes are not made so as to reveal the very smallest and the most distant thing. The utility of a microscope depends upon how plainly it is able to make minute objects visible; and the more a given one magnifies an object, the smaller the portion that can be seen and the less light is available for the purpose, and when the objects are so small as the few thousandths of an inch, the light waves interfere with each other at the edges, and produce colored fringes that cannot be got rid of altogether, and very small objects become indistinct for that reason. Microscope lenses are marked as 1 inch, $\frac{1}{2}$ inch, $\frac{1}{10}$ inch, and so on, meaning by the fraction the approximate distance it must be brought to the object in order that the latter may be seen. The higher the power, the shorter this distance. A one-tenth inch objective may magnify an object a thousand diameters and perhaps more, so that a blood corpuscle having a diameter of only one three-thousandth of an inch may appear about three-tenths of an inch in diameter, and the details of its coarser structure may be very well seen; but if there be a minute point upon it, still indistinct because it is minute, and a still greater magnifying power required to see it, and a $\frac{1}{20}$ objective be taken, the actual magnifying power may be five thousand diameters. But now one is approaching the dimensions of wave lengths themselves, and the agent necessary for observing introduces its own complications, producing distortions and color fringes about the point to be studied, and no way has been found of obviating this. Objectives have been made having a focal length of only the $\frac{1}{50}$ of an inch and one having only the $\frac{1}{75}$, but no work of any importance has ever been done with them. The best of the microscopic work has been done with lenses that magnify no more than one thousand diameters. It is said that the best microscopes will show an object that is no more than about the one hundred-thousandth of an inch in diameter, but it appears simply as a point or a line, and no details of its structure can be seen. Fine rulings upon glass have been made that are known to have this degree of fineness, because the mechanism that rules them can be gauged to that degree; but many persons cannot see these in a microscope, though others can. So within the limits of the visible not a little depends upon the acuteness of vision, and there is a great difference among individuals in this respect. On account of the properties of the ether waves themselves in their relations to each other, it does not appear probable that much improvement is possible to the microscope. This does not imply that we may not know more of the minute structure of bodies than we do now, for there are other sources of knowledge of minute quantities than simply direct eyesight, which are just as reliable, perhaps more so. A good chemical balance will weigh to the millionth part of the load. Whitworth showed that it was possible to measure to the millionth of an inch by touch. The spectroscope will indicate the millionth of a grain by the tint of the gas flame, and the color of a drop of water is appreciably changed by the one three-millionth of a grain of fuschine. Some substances, like essential oils, sulphuretted hydrogen, and the odors of flowers, can be perceived when the quantity is certainly less than the fifty-millionth of a grain.

Any day may bring tidings of new instrumentalities that help in the solutions of the interesting questions concerning molecular structure that are now quite out of our reach. Let it be granted that the problems are altogether physical ones, such as are justified by the known mechanical relations of energy, and one may wait with patience. Let one assume that some or any of them are not mechanical, and he not only is in danger of having to revise his judgment in some degree any day, but he reasons against the significance of all the knowledge we have of matter and its energy.

The larger a lens is the more light can go through it: a lens two feet in diameter will let four times as much light through it as one only one foot in diameter. As remote objects, like the distant stars, appear dim on account of their great distance, it becomes needful to concentrate the light from a much larger area than that of the pupil of the eye. If the pupil be one-tenth of an inch in diameter, a certain amount of light from a star may enter it. A lens one inch in diameter would concentrate at its focus 100 times as much, and one a foot in diameter, 14400 times more; and hence the object would appear so much brighter. Along with this apparent brightening of the star, it is apparently brought nearer and enlarged. There are limits to the size and useful magnifying power of telescopes as well as to those of microscopes. The magnifying power of telescopes depends very largely upon the eye-pieces used, and the shorter their focal length the more do they magnify. The large lens, called the objective, serves mostly to collect a large amount of light. It is to be kept in mind that the movements of bodies are magnified as much as their apparent dimensions, and when there are any movements of the body surveyed, or of the instrument itself, distinct vision becomes correspondingly difficult.

With the telescope the chief trouble comes from movements of the air, which are rarely of uniform quality and motions. Not only its transparency, but its degrees of density caused by heat and wind, are varying all the time; and these seriously interfere with telescopic work. If a magnifying power of say 100 be employed, these disturbing causes are increased in proportion, and with a power of 1000 nothing can be distinctly seen. Suppose, however, the air be in best condition for observations, and a power of 1000 be put upon the moon. As the moon is about 240000 miles away, this magnifying power would have the effect of bringing it 1000 times nearer, or as it would appear to the eye if it were but 240 miles away. Now, an object 240 miles away can reveal no interesting details at all; anything much less than half a mile square could not be distinguished unless it were a very bright or very dark spot. Powers as high as 8000 have been used; and such a one would bring the surface of the moon as it would appear if it were about thirty miles distant, which might show a city, a large town, a lake, and the difference between field and woodland, yet nothing satisfactory was seen for the reasons mentioned, so for most astronomical work a magnifying power of only a few hundred is used; seldom more than five hundred. When large telescopes are set on elevated places like the Lick Telescope on Mt. Hamilton in California, some of the troubles from disturbed air are obviated, and it is hoped something more may be learned about our nearer astronomic neighbors. But these large telescopes collect so much more light that stars so distant as to be quite invisible with smaller glasses become plainly visible with them. With the unaided eye no more than 5000 or 6000 stars can be seen in the whole heavens, with an opera-glass as many as 100000 become visible, while the Lick telescope, with an object-glass three feet in diameter, shows nearly 100, 000000. Each increase in the size of the telescope adds to the number of visible stars, and one cannot but wonder if their number be infinite, or if there be a boundary to the universe of matter. Though the visible boundary of our universe has been greatly extended by the invention of the telescope, nothing has been descried anywhere but matter and motion: there has been nothing added to our knowledge but the sense of bigness. Instead of only a few thousand of hot and flaming stars, there are hundreds of millions of them, made of the same kinds of matter, having the same kinds of motions, controlled by the same laws, and nothing animate in any of them more than in a bowlder in the wall. Clifford said he wished they were farther off. The problems of astronomy are interesting studies in mechanics, but are not inviting to those most interested in life and mind. Herschel and Chalmers and Dick and Mitchell are dead. The knowledge already gained has destroyed both their arguments and hopes, and has left the inhabitants of this earth the possessors of the universe, yet unable to take possession.

If there are inhabitants in Mars they are as unable to traverse space as we are; and the possibility of our yet being able to do that is not half so unlikely as it seemed to be but a very few years ago, since it evidently requires for accomplishment but a directed reaction against the ether; and we already know how to produce the reaction by electrical means; and every point in space has the energy for transformation.

It is generally agreed that the so-called attraction of a magnet for its armature is really due to the pressure of the ether upon the latter, and it may be as great as two hundred pounds to the square inch.

An electro-magnet without an armature is therefore reacted upon by the ether to that degree. When this reaction can in any way be neutralized at one pole and not at the other, the ether reaction will push the magnet backwards, and the navigation of space will at once become mechanically possible.

THE RADIOMETER.

It is a familiar enough fact that when sunshine falls upon a surface the latter becomes heated. In general, the darker the color of the surface the more rapidly does the temperature rise; and some bodies, when thus exposed for some time, become unbearably hot. We are able to say that the surface molecules of such a body are in a brisk vibratory movement; that they have more energy than other bodies with less temperature. If one imagines the condition of things when the molecules of the air impinge upon such a heated surface, he will understand how they must bound away from it with greater velocity than they struck it with, and if with greater velocity, then with greater energy. As action and reaction are equal, it must kick back upon the surface as it leaves it, thus tending to make the surface move in the opposite direction; and a large number of

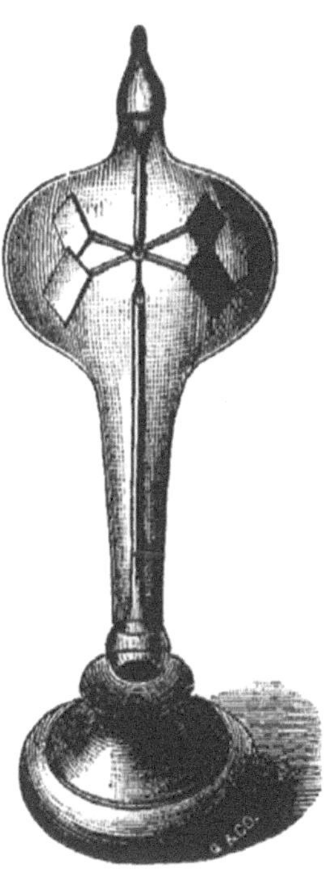

DIAG. 12.—RADIOMETER.

such impacts must give a resultant backward pressure. If the surface be a small one, the increased pressure in the air in front will travel round to the other side at the rate of eleven hundred feet in a second in ordinary air; so the pressure will be equalized in a very short interval of time. If the air be rarefied in front of such surface to such a degree that the free path of the molecule is many times greater than its ordinary length, that pressure cannot get round nearly so fast, and there will consequently be a constant backward pressure, produced by the molecules that impinge upon it and become heated by contact with it. The pressure per square inch is very slight, as it is produced by a relatively small number of molecules; but it may be made apparent by mounting some disks, blackened on one side, upon a pivot in a glass bulb, and, after exhausting a large part of the air, hermetically sealing the bulb. Such a device is called a radiometer. When put where sunshine, or the light from the flame of a lamp or candle, or even the heat of the hand, may fall upon it, the vanes begin to rotate, the blackened side backing away from the source of the energy. This movement was at first interpreted as being due to the actual pressure produced by light waves, but further investigation showed that idea to be wrong. The movement comes from the transformation of the motions of ether waves, first into heat, and second into the translational mass motion observed. The radiometer is, therefore, a machine for transforming ether waves into visible mechanical motions.

PHOTOGRAPHY.

It has already been explained how heat acts upon molecules, increasing the amplitude of the vibrations of the atoms that make them up, and, if carried to a sufficient degree, is able to quite destroy the molecular structure and enable the component atoms to enter into new combinations. The degree needed for this depends upon the kind of molecules. Some molecules are so stable that only the very highest temperature we can produce can break them up. Others are so feebly cohesive that the least touch will cause them to go to pieces, and sometimes with explosive violence, as is the case with what are called fulminates, compounds of nitrogen with silver or with mercury; and sometimes the same result is reached by ether waves, whose number per second is such as to set one of the ingredients into sympathetic vibration and thus decompose the compound, doing it at a slower rate than the others. Nearly all complex molecules are decomposable in this way, and the process is going on all the time in nature where there are organic things to act upon, but the process is usually slow.

When shingles are first laid they have a fresh surface and new appearance, which is presently lost by the exposure. Take a freshly planed piece of soft pine or other white wood, and fasten to the surface a piece of paper cut into any shape or design,—a circle, a star, or the like,—and set the wood where the sun can shine on it for a few days. When the design is removed the figure will be plainly seen on the wood by the difference in tint between its surface and that part which the sun has shone upon. The latter is much darker. This is an example of photographic action, as is the color of fruit, etc.; for if a design is pasted upon a green apple, which is red when ripe, the design will protect the surface from the action of the light, and will therefore appear upon the apple in a light tint. Diagrams and letters may be fixed thus upon fruit of any kind. Discolorations of all sorts, due to ether waves or light, may properly be called photographic action, both fading and darkening, as when the skin becomes tanned. For practical purposes some compounds of silver are generally employed, because they are more sensitive to the action of visible waves than most other substances. They have the property of being easily disorganized by waves whose length ranges from about one forty-five-thousandth of an inch to those in the neighborhood of the seventy-thousandth of an inch, some of these being visible waves, the others being too short for visibility. When a surface is prepared with some one of the sensitive salts of silver,—generally the iodide or the bromide,—and a picture of an object produced by the lenses of the camera is allowed to fall upon it, the decomposing action is proportional to the amount of light and shade in the different parts; and, when the plate thus exposed is placed in certain chemical solutions called developers, the decomposition is completed and the products dissolved out, leaving a coating of pure silver, with a thickness proportional to the chemical action that has taken place. This gives, then, a correct likeness of the picture that was in the camera. Formerly it took a long time to produce such a picture, a person having to sit still for half an hour or more. More and more sensitive preparations were produced, until now a good picture can be taken in less than the thousandth of a second; and the practice of the art has become a great industry. There are many preparations in common use for taking such pictures, but nearly all of them have silver for the chief constituent. It may be remarked that silver compounds are remarkably unstable. Silver is not easily oxydized, for it remains untarnished for an indefinite time, as exemplified by coins and jewelry. But there are plenty of other compounds that may be used. Thus the common blue-print is a compound of iron. The salts of chromium are also sensitive to such waves.

It was remarked that the salts of silver are sensitive to ether waves between quite a wide range in wave lengths, but the longest of them is in about the middle of the visible spectrum. They reach from there into the region beyond the violet. Yellow and red waves are incapable of affecting such a preparation, while the waves that are the most efficient for it are the blue ones.

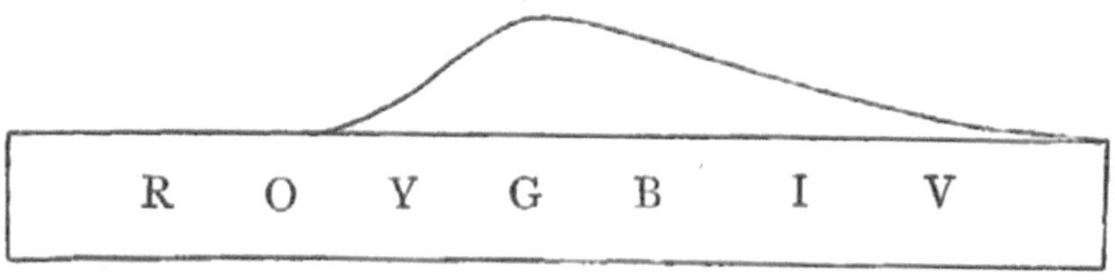

DIAG. 13.—PHOTOGRAPHIC RANGE FOR SILVER SALTS.

Other substances have a different range, and a curious chemical discovery has shown that silver molecules may be loaded; that is, may have attached to them in a temporary way some other kinds of molecules that render them sensitive to waves of any length. If an ordinary photographic plate has a solar spectrum thrown upon it, there will be no indication of action below the green; but, if aniline be added to the sensitive coating and the plate be then exposed in the same way, the action will now be seen to have gone on to a distance below even the longest red wave that can be seen. In this way photography has shown that the spectrum of most incandescent bodies is much longer than the visible part of it in both directions. It was the observation that photographic action took place most strongly in the blue part of the solar spectrum, and in the region beyond, that led to the belief that light waves and chemical rays were, in some way, unlike each other. From what has been said it will be seen that the reason for the different action was due to the character of the material used. When a molecule is made bigger or heavier in any way, longer waves can affect it more; and that is the significance of the so-called loaded molecule. In reality, the whole molecule is made more complex and bigger, and longer waves can shake its atoms loose.

It is to be hoped that all can understand that there is nothing mysterious about photographic action; that it is as simple in its mechanical principles as anything can be. One may not be able at once to say in any given case which atoms or which parts of a molecule are loosened by the vibratory strains. In this one it may be the nitrogen, in another it may be the silver, and in still a third it may be oxygen; but in each case the mode of action is the same, and it may be said to be mechanical throughout.

VISION.

Our various senses differ much in their mode of action, and require for excitation not only each its proper stimulant, but degrees of remoteness from actual contact to the most distant points. Thus the sense of touch requires absolute contact of a body: so also does taste,—the sugar or the salt must dissolve upon the tongue. A distance of but the tenth of an inch between the sugar and the tongue will be absolutely prohibitive to the consciousness of sweetness. The sense of smell requires the actual contact of the gaseous molecules upon the nasal membrane, but currents of air and gaseous diffusion secure to us this condition, so that the emanating body itself may be at some distance, and yet we become conscious of the bank of violets, the cup of coffee, or the chemical laboratory. This sense, therefore, enlarges our field, so to speak, and permits us to be conscious of bodies out of our immediate reach. The sense of sound still farther enlarges the space that can react upon us. But the loudest sounds, such as the roar of cannon and thunder, lose their intensity shortly, and can rarely be heard beyond a few miles. If our endowment of senses stopped with these, we should really be quite limited in our possible knowledge; for as we can know only what comes into our experience, how small the possibilities of existence would be to us! What we could touch, taste, smell, and hear we could know something about, though we were unconscious of any lacking sense. We should need some apparatus that could make us conscious of the most distant things as well as those close at hand. We should need just what we have got,—the sense of sight, that extends the field of experience and of interest to us to the boundaries of creation. The other senses give us information of contiguous things, but sight brings the universe itself to our consciousness.

The sense of touch is diffused all over our bodies. There is no such thing as an organ of touch. The senses of taste and smell are restricted to localities and to organs that have other functions as well. Only sound and sight have specific organs, having no other function than to respond to sonorous and optical motions, and thus they have a peculiar dignity in the physiological mechanism; and precisely because the eye and ear have these mechanical functions do they come into the domain of physics. They are machines by which certain forms of motion are transformed into others suitable for nerve transmission to the seat of consciousness.

It has often been pointed out that the structure of the eye is like the camera of the photographer. In each there is a chamber *a*, having a lens in front, which has a length of focus adapted to the distance

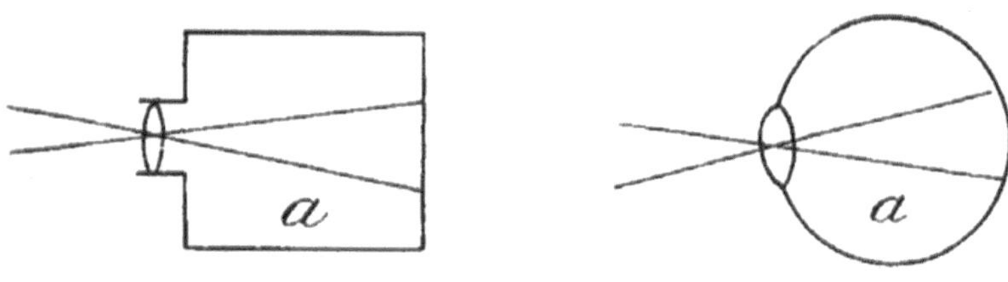

DIAG. 14.

between it and the back of the chamber, so that the image of objects external to it will be produced by it upon the back of the chamber, where there is in each a sensitive coating so affected by the light as to make an impress. In the camera this action has been explained as chemical reaction when molecular dissociation results, proportionate to the amount of light that falls upon any part of the surface exposed.

In each there is an arrangement for altering the focal distance of the lens. In the camera it is a ratchet-wheel that moves the lens towards or away from the back.

In the eye there are muscles attached to the edge of the lens that by contracting make the pliable lens less convex and so increase its focal length. For the camera there is an exchangeable diaphragm having perforations of various sizes to admit more or less light through the lens. In the eye there is a colored muscular disk called the iris, that contracts or expands in an automatic way so as to expose more or less of the lens to the light. The functions of the two devices are identical.

DIAG. 15.—PHOTOGRAPHIC CAMERA.

The energy possessed by the ether waves that fall upon the sensitive photographic plate is spent in doing the molecular work of disintegration. In the eye all the energy is stopped at the sensitive back coating called the retina, and must of course be accounted for in some physical way. In the camera all the energy of the waves is spent in precisely the same kind of a way; that is, there is no such distinction as what is called color in it: and color

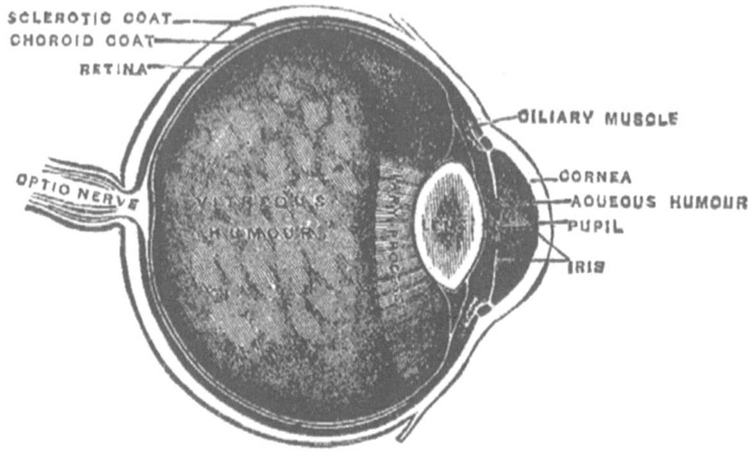

DIAG. 16.

photography—that is, the direct picture of objects in their proper, natural tints, such as we know they have, upon the sensitive plate—has not been accomplished, for the probable reason that the colors of the molecules that are the result of the decomposition of the silver compound are either transparent or blueish black. In the eye, the distinction between wave lengths which we denominate color sensation is very pronounced.

The sensations are so much complicated with the processes that induce them that it is not always easy to keep in mind the purely physical side or the subjective side while treating of them.

The following are some of the more common phenomena of vision which must be taken into account in forming any judgment or theory of it.

When a firebrand is swung round and round it leaves an apparent luminous trail, the length of which depends upon the rapidity of motion. This is called the persistence of vision, and indicates that the sensation does not cease instantly after the source has gone. If the brand be swung round at the uniform rate of once per second, the length of this luminous trail will be a rough measure of the duration of the sensation after it is once excited. Thus, if it appeared to be one-quarter of the circle, the sensation must last for one-fourth a second. For impressions not very bright the sensation lasts but about the tenth of a second. If, however, the object looked at be very bright, like the sun, for an instant, the sensation may last for many seconds; and, in general, the older the person the longer does it last.

Different colors also have different degrees of persistence. Violet, blue, and green soonest fade out, and red is the last to vanish for most eyes. This signifies that wave length has in someway to do with the persistence. When a bright colored object, like a bit of red paper, is put upon a sheet of white paper and steadily looked at for a few seconds, and is then suddenly removed while the eyes are kept fixed upon the same place, the image of the red paper will still be seen, but it will appear with a green tint, and will fade out in a few seconds. A green piece of paper, or any green object looked at in the same manner, will give an image in red. Blue ones give yellow, and yellow blue; and these tints seen in this way are called complementary to each other, as it is found by combining such together they produce the sensation of white light. Whiteness is therefore a compound sensation. Formerly, it was thought that white was only produced by the composition of all the colors of the spectrum in the same proportions they exist in it; but the same sensation of whiteness can be produced by red, green, and violet, and by blue and yellow. This is not to be understood as applying to pigments or paints, but to light itself.

If one looks at a strongly lighted object intently for a few seconds, and then turns his eyes to a dimly lighted drab surface, he will be able to see, sometimes in a surprisingly realistic way, the same object against the new background. If it be a person looked at, the features may even appear in a startling way. The size of the subjective figure will depend upon the distance of the background, being larger the more remote that is. Age and health have much to do with the persistence of such sensations. Young and vigorous persons seldom notice them until they carefully look for them; while older ones, and especially weakened ones, may be much troubled by them. Some nervous systems react upon the eye itself, and give rise to similar images there; and these subjective images have not unfrequently been mistaken for objective persons living or dead. The color a given object appears to have is not unfrequently modified by what colors the eye has been resting upon the instant before, and hence two persons may look at once upon the same picture and see it in very different tints.

As ether waves are the source of the sensation, it is obvious that a certain number of consecutive waves must be necessary to affect the eye; that is to say, it is not in the least probable that a single wave of any length could produce a sensation. How many are needed is not known, but one can determine somewhere near what the number must be if he knows how brief a time is sufficient to produce a sensation. It is said that some flashes of lightning have been found to occur in less than a millionth of a second, and those may produce a very strong sensation.

If there are five hundred million million vibrations per second, as we know there must be to give such a sensation, in the millionth of a second there must be five hundred millions; if the brightness were reduced ten thousand times and it were still visible, there must then have been not less than fifty thousand waves: and this is equivalent to saying that the eye could perceive light if it lasted no longer than the ten thousand millionth part of a second, which is probably true. But there is another condition; namely, the *energy* of the waves must be sufficient to effect a physical change in the eye; and we know that the energy of such ether waves varies with the square of their amplitude. If, then, any wave whatever has not energy sufficient to produce the necessary physical disturbance in the eye, it could not produce vision. And this is the most probable reason that we do not see in what we now call darkness. It has been shown that all matter at all temperatures is vibrating and setting up ether waves, and also that in all liquids as well as solid bodies there are vibrations due to their atomic and molecular interference; and, theoretically, there must be vibrations of all wave lengths at all times and in all places, but at low temperatures the shorter waves, though not absent, would have but small energy, and, as the body becomes hot and the shorter ones acquire more, it is done at the expense of the energy of the longer ones, for the light given out by an incandescent lamp increases faster than the supply of energy to produce it. It therefore appears as a necessary conclusion that the reason we cannot see in the dark is not so much because the waves of proper wave length are entirely absent, as that they have too little energy to affect our eyes. Other animals, such as rats, mice, owls, bats, and the like, can see where it appears to us to be pitch dark. They must, therefore, have eyes adapted for longer wave lengths than are ours, or else the sensitiveness of their eyes exceeds ours. As they see readily in the daylight, it is certain they are adapted to such waves as our eyes are; and, if ours were sufficiently sensitive, or had a greater range in effective wave lengths, there would be no such condition as darkness. That is the same as saying that darkness is in us rather than being a condition external to us.

THE THEORY OF VISION.

When it was discovered that the sensation of whiteness could be produced by combining three different colors,—red, green, and violet,—it was inferred that there were probably three sets of nerves that were spread as a fine net-work over the retina so that either of these rays might fall at any point in the field of vision upon it and so produce the sensation. At the same time, when one or two of them were absent, the other nerve ingredient would be present to be affected;

and, furthermore, each one of these three nerves was sensitive to quite a wide range of wave lengths, and their overlappings gave perception without any break from the extreme red to the extreme violet. In this way color perception could be explained. This view was adopted as a working hypothesis; and there was no other proposed, although there was no evidence whatever for the existence of three sets of nerves having different properties. It has, however, lately been discovered that the retina secretes a substance called purpurine, on account of its purple tint, which is very rapidly bleached or decomposed by the action of light. That is to say, it possesses photographic properties in a marked degree. This discovery has led to the view that vision may be altogether due to photographic action, and the older view has been about abandoned. The details of this theory have not yet been all worked out, but the purport of it may be briefly stated.

Given the purpurine spread over the retina: this would be its sensitive coating corresponding to the silver preparation upon the photographic plate. The action of the light upon it being the same in character, decomposes it into simpler molecular compounds. The optic nerve is certainly spread over the retina, and the purpurine is in its meshes, and any disturbance taking place in this substance must correspondingly affect the ends of the nerves imbedded in it. Given the disturbance that can affect the optic nerves, and it is transmitted at once to the base of the brain and there interpreted as light sensation. The differences there might be in the amount of disturbance would be the differences that are called brightness or intensity. If molecules are disintegrated, as in photographic action, there must be a relatively large amount of free-path motion resulting from the wave action in the eye, and the amount of it proportional to the energy expended. Such an effect would give a general sensation of light, probably, also, effects of light and shade, so the forms of bodies would be readily enough seen. It would also account for persistent effects; for, when molecules are made to move fast or slow, they do not cease instantly on the removal of the source of the motion, but they continue to thus move until their energy has been reduced to that of the surrounding medium. With simple purpurine there appears to be no more possibility of chromatic effects than there is in the common silver preparation on the photographic plate. Suppose, however, the purpurine to be not a simple kind of a body, or made up of only a single kind of molecules, but instead made up of as many as three different kinds having as many different molecular weights, and, therefore, capable of being reacted upon by three different wave lengths.

Call these three substances *a*, *b*, and *c* purpurine. Let *a* be such as red waves can decompose, *b* such as green ones can decompose, and *c* such as only the short purple ones can break up or shake up. If these are uniformly mixed together and spread over the eye, then red waves would shake up the red constituent, but would leave the others alone; and the same would hold true of the others. If one has been looking at red-light wave lengths, the *a* purpurine would be used up, but the *b* and *c* would still be present unimpaired; and now, when white light is again looked at, the *b* and *c* would be acted on strongly because they are present in greater quantity. The resulting sensation would be the compound of these two reactions, which, as is well known, is a greenish tint. In a like manner, each of the others when used up would leave the same field fresh with the other constituents, and so give the complementary tints; and in this way chromatic effects of all sorts can be accounted for.

Some persons are color-blind; that is, they are unable to distinguish some colors; and this defect is usually for red rays. Such a color-blind person will be unable to see the red end of the spectrum, and the colors of it will appear to leave off in the yellow or orange. The old explanation was that the red sensation nerves were absent. The newer explanation is that the *a* ingredient of the purpurine is wanting either partially or altogether.

Of course it is to be understood that the products of decomposition by light in the eye are removed and fresh material secreted in its place by the organ itself in a manner similar to the removal of waste tissue and its repair in any other part of the system.

The function of the retina, then, would appear to be the secretion of the sensitive substance needed for vision, instead of itself being the sensitive substance.

Such an explanation of vision makes the eye still more like the photographic camera than appears in its outward form and mechanical functions. And thus one is able to trace the forms of motion that constitute the heat and the temperature of a body through its resultant ether waves to the molecular break-ups at the ends of nerve fibres, whence the characteristic motions are transmitted to the base of the brain, to be interpreted thus or thus, according to position, number, and energy. We begin with motion, we end with motion at the seat of consciousness, and there we stop. It is vibratory in the hot body it starts from, it is undulatory motion in the ether, it is oscillatory in the disrupted molecules, and a longitudinal wave in the nerve. Whether it is discharged from further service at the base of the brain, or is stored up in some way as experience, no one can say; but it is certain that a relatively large amount of molecular energy finds its way constantly to the brain, and some of it is re-employed as reflex action, giving rise to voluntary and involuntary muscular and secretory processes, as when one winks, or dodges a threatening motion before the will can act, or laughs or weeps at sights and sounds. In either case the result is the physical expression of a physical antecedent, with an intermediate mental quality called emotion.

The eye may then be said to be a machine for the transformation of ether waves into interpretable molecular or atomic motions, and its function ceases at the ends of the optic nerve.

CHAPTER VIII

Electricity

THE industrial applications of electricity are now so extensive and varied that every one is acquainted with them in some measure, and yet fifteen years ago there were millions of persons in the civilized nations who had never seen an electrical phenomenon with the exception of lightning. The apparently capricious behavior of lightning, together with the attractions and repulsions exhibited by electrified bodies, were phenomena so different in character from any other, that it came to be looked upon as a very mysterious force. Fifty and more years ago it was classed with heat and light as one of the imponderables. To-day even the question is often asked, What *is* electricity? with the emphasis on the word "is," as if one knowing enough might describe it as he might describe a genii or an object having specific qualities that might be isolated from everything else. Some have thought it to be a fluid, some two fluids, some vibratory molecular motion, some a property of matter, some a motion in the ether, some the ether itself; and, lastly, some have concluded that we do not and never can know its nature. Hence, to-day there is no generally received notion concerning its nature.

Still, one may know a great deal about the agent itself,—how it originates, what it will do, and its relations to other phenomena,—and not concern himself at all as to the nature of it. Heat and many of its laws were well known before any one knew or even suspected what its nature was. The law of gravitation is known and applied on the scale of the universe without demanding any explanation of the phenomena, and it is equally true that our knowledge of electricity is very extensive and accurate, and doubtless what we do not know to-day we may know to-morrow.

ORIGIN OF ELECTRICITY.

It is here to be assumed as known, that various instruments, such as electrometers and galvanometers, are employed to detect the presence of electricity, and descriptions of them will not be given. Attention will be paid chiefly to the conditions that are present when electricity is generated.

1. THERMAL ORIGIN.

When two different metals, such for instance as copper and iron, are touched together, they are found to be electrified; that is, an electrometer shows the presence of electricity. A piece of copper wire twisted to a piece of iron wire always becomes thus affected, but the effect is so slight that only delicate and sensitive apparatus will detect it. Wires of any of the metals under similar circumstances exhibit the same phenomenon, but in different degrees. This electrification is but transient; in a few seconds it has vanished. If the junction of the metals is heated by the fingers, or in any other way, the electrical condition is maintained indefinitely. If one will imagine such a compound wire bent into a ring so the ends nearly touch each other, it could be shown that the ends attract each other, the attraction being but slight. Here we are not so much concerned about the measure of what is taking place as with its character. If the ends of the wires be allowed to touch, and the twisted junction be kept warm, a current of electricity will continue to circulate through the ring; and, if the ends be connected to a galvanometer of sufficient delicacy, the needle would be continuously deflected, so long as the junction was warmer than the outer ends of the wires; and the deflection of the needle would be found to vary with the difference in temperature between the inner and outer junctions. Some metals, such as bismuth and antimony, when fastened by solder, or in any other way, give much stronger effects with a given temperature at their junction. Such a combination is called a thermo-electric pair. By joining a number of such together, so that alternate ends may be heated at once, the electrical effect is increased proportionally: two will give twice, and ten ten times as much, and so on. When a number of these are nicely compacted together and provided with binding-screws, they are called thermo-electric piles, and are of service in some investigations. It is not necessary, however, to have two different metals in contact to obtain the same kind of effects. If a piece of soft iron or platinum wire be wound into a close coil about a lead-pencil and the ends of it connected to a galvanometer, a current of electricity will traverse the circuit when one end of the coil is heated in a flame. If the other end be heated, the current will go in the opposite direction. The twisting of the wire into the coil produces a strain among the molecules that changes the physical properties to a slight extent: the density is altered. It therefore appears that in this case, as in the cases with two different substances, we have two *physically* different bodies, though of the same element. The facts may be generalized by saying that, whenever two differently constituted bodies are placed in contact with each other, electricity is generated, and is maintained so long as there is a difference in temperature between the junction and the external ends.

DIAG. 17.—THERMOPILE.

If one inquires for the origin of such manifestation as the first case, when two different metals are placed in contact, attention must be directed to the actual molecular condition of the two metals. Suppose them to have the same temperature,—as they have different atomic weights their vibratory rates cannot be the same,—and when the surfaces are put in contact there must be a re-adjustment of their molecular motions, for each will interfere with the other. This disturbance of molecular rates is a disturbance in their relations of energy, and furnishes the energy for the electrical phenomenon that ensues. When equilibrium is restored, as it may be shortly, there is no longer any electrical exhibit.

When heat is applied so as to keep the junction continually hotter than the other parts, the first effect is continuous; for as each element has its own proper vibratory molecular rate, which is increased by the heat, the interference is kept up and an electrical current results, which the heat is spent to produce and maintain. One needs to have in mind what is signified by heat as vibratory atomic, and molecular motion, in order to clearly perceive what is expended in the thermo-electric pile. The face of the pile, when it is generating a current of electricity, does not acquire that temperature it would acquire if it was prevented from producing a current by having the wires detached. Hence the amplitude of vibrations is lessened by the electrical work done, and we may say that heat has been converted into electricity, a thermal origin.

2. CHEMICAL ORIGIN.

When a piece of copper is dipped into a vessel of water, and a wire leading from it is connected to a proper electrometer, it is found to be electrified to a certain degree. If a piece of zinc be substituted for the copper, it too indicates a still greater degree; and now let both be placed in the same water and connected by a wire, and a current of electricity will flow through the wire, as in the case with the thermopile. This current will be a transient one, or very slight, if the water be pure; but if a little acid like sulphuric be added to the water, the current may be relatively a strong one. If, instead of the zinc and copper, any other two metals be taken, the results will differ from the former only in degree. Zinc and copper, or zinc and carbon, are generally employed, because those have been found to give better results than other available elements; and such a combination of metals, with some solution, acid or alkaline, which is capable of dissolving one or both of the metals, is called a galvanic battery. A single jar with its proper elements is called a cell; and by the addition of cells additional effects may be produced; that is, with two cells twice, and with ten cells ten times the effect.

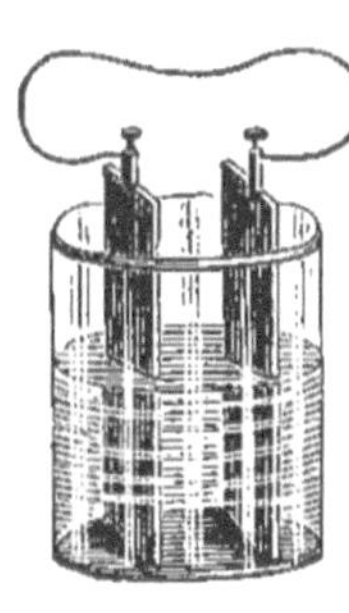

DIAG. 18.
GALVANIC CELL.

As with the thermo-pair, one may inquire what conditions were known to be present that could furnish an antecedent to the electrical current that results. This is answered by pointing out, as in the other case, that there are two substances differing in their physical qualities, copper and water, or zinc and water, and molecular rearrangement at their junction must necessarily result. More than this. It is known that the zinc and oxygen have a strong affinity for each other. The oxygen is combined with hydrogen to form the water, and in water the molecules are without any definite arrangement: they face in all directions, and move about with the greatest freedom, with but little, if any friction. When zinc is placed in it, the attraction of the zinc for the oxygen part of the molecule must result in making every water molecule in proximity to the zinc swing round so as to present its oxygen side to it. This orientation of the liquid molecules is called their polarization. The attraction between the two is not quite strong enough to disrupt the water molecule; but the addition of sulphuric acid weakens the attraction between the hydrogen and oxygen, and enables the oxygen to seize a zinc atom, and both combine with the sulphuric acid to form the sulphate of zinc. Here we have chemical reactions such as always result in exchange of energy; for the sulphate of zinc has less molecular energy than the zinc, the water, and the acid, in the same way that carbonic acid gas has less energy than the carbon and oxygen gas that formed it. There has been, then, a molecular change accompanied by the development, first of heat and second the generation of electricity; for if the electrical current be not allowed to flow, the battery cell will itself heat up more than it otherwise would do. There are chemical, thermal, mechanical, and electrical phenomena here, which may be perceived by carefully thinking of the successive steps in the process. The distinctive thing here is to bear in mind what the characteristic antecedents of the electrical phenomena are. What are the chemical, the thermal, the mechanical factors, except special forms of exchangeable molecular motions? So one may say that in a galvanic battery chemism or heat has been transformed into electricity. Though the mechanism of transformation is different, yet the same factors appear as in the thermopile.

3. MECHANICAL ORIGIN.

When a piece of glass or of wax is rubbed with a cloth or catskin, the two substances subject to the friction become endowed with a new property which they do not otherwise exhibit. If a glass disk be mounted so as to be rotated, and proper connections made to it, as in the common Static Electrical machine, a

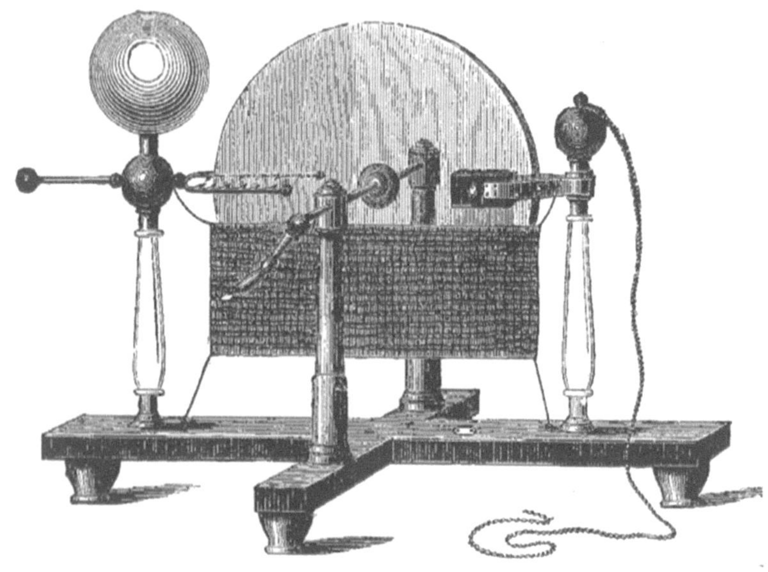

DIAG. 19.—STATIC ELECTRICAL MACHINE.

current of electricity may be maintained by maintaining the friction, and all the electrical phenomena may be produced that can be with electricity from any other source. They are identical, but the source is the friction of dissimilar substances. It will be recalled that dissimilarity in substance was the condition in each of the former cases; but in this, mechanical friction is the second factor. In the chapter on heat it was pointed out, and it is a familiar enough fact everywhere, that heat is always the immediate result of friction. So in this mechanical source, with apparatus so dissimilar in all outward form to both thermopile and galvanic battery, we still have precisely the same molecular conditions that were operative in them to produce electricity,—two dissimilar substances, and heat or a kind of motion that results at once in heat.

4. MAGNETIC ORIGIN.

If a wire of any sort be placed across the pole of a magnet, and held quiet there, no electrical effect will be noted; but if the wire be moved toward or away from the pole, it will become electrified, and if one end of it be connected to an electrometer the movement of the needle will indicate it. If the two ends of the wire be connected to a galvanometer, whenever the wire is thus moved in front of the magnet pole a current will flow through the circuit, and the movement of the needle this way or that will indicate the motion of approach or recession. The strength of this current will vary with the rapidity of the motion of translation of the wire through the space in front of the magnet; and the wire through which it goes becomes heated. This is the same as saying that the mechanical motion of translation of the wire is converted into heat in a manner as if it had been subject to ordinary friction there; and as a matter of fact, it is found to require more energy to move the wire in such a space when the ends of the wire are in contact, than it does when they are not. This shows that the material of the wire is subject to some restraint under such conditions and in such positions, and the degree of restraint depends upon the distance it is from the magnet, as well as upon the strength of the magnet itself. Hence the different parts of the wire are in different physical states. And this is just what is exhibited by the twisted wire in the case of the thermal origin; and when motion is imparted to the wire, the degrees of stress in it change, and a current of electricity is the result. That such a stress is really present in the wire can be proved in several ways, which only need to be alluded to in this place. First, the electrical resistance of a wire is greater when in front of a magnet than elsewhere; and second, the phenomenon known as Hall's, in which a current of electricity going through a conductor is deflected from its course in the neighborhood of a magnet. So we have, in this magnetic origin, two bodies with different physical constitutions and external motions impressed upon them, which gives the electrical product observed.

5. ELECTRICAL ORIGIN.

Imagine two wires parallel to each other and a foot apart. If an electrical current from any source is made to traverse one of them, a corresponding current will be initiated in the other, but in the contrary direction. In a like manner if a constant current be kept in one of the wires, and the other one be moved towards and away from the other, currents will be set up in it. Their direction will depend upon whether the motion be approach or recession. The effect is the same whether either or both move at the same time. The effect is similar to the one described under the head of Magnetic Origin, showing that in some way the space about a wire having a current of electricity in it is substantially similar to that about a magnet. The process is called electro-magnetic induction in both cases, and the explanation is the same in this as in the other. It will be well, however, to point out that there are steps in this process that need attention for the sake of mechanical clearness.

Given say, an electro-magnet, through which a current can be sent at will, and so be made magnetic, and with the wire in front of it as before. There is now no magnetism and no electricity in the wire. Make the iron magnetic, and the current is at once induced in the other. I say at once, but this does not mean instantaneously. It takes a short time for the effect of the magnet upon the ether to travel to the wire and affect it. As no electricity escapes from the electro-magnetic circuit, the electricity observed in the wire, or second circuit, is generated in it, and the *immediate* antecedent of it was the stress in the ether which was produced by the magnet. Hence an electrical current can arise from a proper kind of stress in the ether, no matter how that is produced, as one of the factors; the other factor being motion of some sort, mechanical or otherwise. The steps are, an electric current in a conductor, an electro-magnetic effect of the current upon the ether, the reaction of the ether upon the second conductor. Let these steps be kept in mind always when thinking about inductive action, and there can then be no confusion from trying to imagine how electricity gets from one circuit to another when they are insulated from each other.

6. PHYSIOLOGICAL ORIGIN.

There are certain kinds of fish that are capable of giving powerful electrical shocks to men and animals. They are provided with special organs for this purpose, but they have not been the subject of much study for several good reasons. First, they are only to be found in a few localities, and are difficult to obtain; and second, their electrical qualities cannot be studied except when they are alive; and when they are living and healthy their shocks can kill both men and animals, and few are willing to incur the risk. Both mankind and animals in general can give rise to electrical currents. By grasping with the thumb and finger of both hands the terminal wires from a delicate galvanometer, a current is indicated,—a part often due to thermo-electric action, and a part to physiological action,—and it will vary with the tightness of the squeeze of contact and the person experimenting, some developing much more relatively than others. It also varies with the parts of the body in contact with the wires. This physiological effect is always extremely minute, and is not to be mentioned beside the amount necessary to effect the remarkable things said to be done by personal electricity, such as moving chairs, tables, etc. I do not think any one has been found whose physiological electricity could do so much as raise a grain the tenth of an inch.

The various processes continually going on in the body, such as breathing, digestion, blood-circulation, and muscular motions of all sorts, and under conditions of different temperatures, different material, different chemical reactions, are quite sufficient to account for all that has been observed in this direction.

7. ATMOSPHERIC ORIGIN.

The origin of lightning, so far as details go, has never been satisfactorily accounted for. It is obviously not an affair that can be investigated in any very scientific manner, for one can never control any of the conditions when it arises.

Some have thought it due to the condensation of electrified vapor molecules condensing into drops of water, the degree of electrification increasing with the size of the drops. How the original electrification of the molecules was produced is not explained by such. There is no doubt but that a large amount of energy is often involved in a stroke of lightning, judged by its sudden destructive work. The immediate source of this energy is the question. There is no doubt but when a gas or a vapor is condensed into a liquid, a notable amount of energy is liberated in motions of some sort; for it requires energy to be spent upon water to produce the vapor. This is given back when the process is reversed. This energy has often been called latent heat. If this process goes on faster than it can be conducted away, it must either be transformed, or the process must stop. We know, too, that heat motions are most freely transformed into electrical by the phenomena of the thermopile and the galvanic battery, and it is not improbable that this is the source of the atmospheric electricity. It is certain that where it originates there are two differently constituted kinds of matter,—the air and the water; and it is equally certain that there are some vigorous exchanges of motion, both in the form of wind and heat, and these are the conditions present in each of the cases where our knowledge is most complete.

One may then fairly conclude from the analysis of all the known sources of electrical development, that motion of some sort is the antecedent in every case. This motion may be the sort called mechanical, or that called molecular or atomic, as heat, but it is always a factor; and the amount of electrical energy developed in every case is equal to the *immediate* mechanical, chemical, or thermal energy which disappears when it is produced. If one admits that the quantity of energy in phenomena is constant, that the quantity of matter is constant, there is but one variable factor, and that is motion. If mechanical motion is transformable into heat, and heat into electricity, and some known form of motion is the invariable antecedent to the production of electricity, it does not need a very profound logician to say, *so far*, the nature of electricity is known.

ELECTRICAL TERMINOLOGY.

Every particular science and art has some technical terms to give precision and definiteness to its processes and its laws, and the advances made in any science depend very largely upon exact signification of its terms. The late rapid development of electrical science is due in a large measure to terminology, adopted about twenty-five years ago; for it enables a man not only to know what he himself is talking about, but also to understand others, and that was not the case before. A system of units and names for them are matters of the first importance. How these were derived need not be stated here, but it is needful for every one now to understand the significance of the more common of them.

Imagine a wire in front of you with an electrical current traversing it from left to right. If it travels in that direction it is because the electrical pressure is less towards the right than in the opposite direction, just as water flowing through a pipe towards the right travels thus because gravitative pressure is less in that direction than in the other. Gravitative pressure is measured in pounds, electrical pressure is measured in *volts*.

If the pressure at the left of the wire were increased

in any way, there would be an increased current of electricity in the wire, just as there would be more water go through the pipe if the head or gravitative pressure were increased. The rate of water flow might be measured as so many cubic inches or cubic feet per second. The rate of electrical flow is measured in *ampères*.

If the water pipe were a large one, and the pressure the same, more water would flow through it than if it were a small pipe of the same length. In like manner a thick wire will permit more electricity to flow through it with a given electrical pressure than a thin one. The water pipe is said to oppose friction to the movement of water.

A conductor of electricity is said to offer resistance to the flow of electricity. No name has been given to any unit of frictional resistance, but electrical resistance is measured in *ohms*.

A definite quantity of water flowing at a given rate will be emptied from the pipe in a second or a minute. So will a definite quantity of electricity go through the wire in a second or a minute. The quantity of water thus flowing would be measured as so many cubic feet, or so many gallons; the quantity of electricity is measured in *coulombs*, a coulomb being an ampère per second. Where the rate of flow of an electrical current is given in ampères, the quantity will be found by multiplying the ampères by the number of seconds the flow has continued. Thus a ten ampère current in an hour will have conveyed $10 \times 60 \times 60 = 36000$ coulombs.

There are also measures of capacity. The cubic inch, the cubic foot, the pint, quart, bushel, and so on, are measures of volume or capacity: any of them may be adopted as a unit, and when accuracy is required all are reducible to the cubic inch as a standard. Thus in a gallon there are 231 cubic inches.

In electricity the unit of capacity is called a *farad,* and it represents the capability of an electrical device to receive and hold a definite amount of electricity under the standard conditions of pressure. Thus, when under a pressure of one volt it holds one coulomb, the capacity of the apparatus is said to be one farad. Actually a piece of apparatus of sufficient size to hold that quantity has to be so enormously large that a much smaller one was requisite for convenience, and consequently the microfarad, or the one-millionth of the farad, has been more generally adopted.

As work may be got out of a flow of water, the amount of work depending upon the pressure and the rate of flow, so may work be got from an electric current, the amount depending upon the pressure, volts, and the current, ampères. The product of these factors, volts into ampères, is called *watts*; and the mechanical value of one watt is such that 746 is equal to a horse-power, which, as before stated, is 550 foot pounds per second. The working power of a watt is therefore $\frac{550}{746} = .735$ of a foot pound per second.

OHM'S LAW.

This is simply that the current in an electric circuit may be determined by dividing the electric pressure in volts by the resistance in ohms. It is customary to use symbols for each of these factors, E or E.M.F. (electro-motive force) for the pressure in volts, R for the resistance in ohms, and C for the current in ampères, so Ohm's Law when thus written reads $\frac{E}{R} = \mathrm{C}$. Recurring to the idea of a wire in front carrying a current of electricity from the left to the right, and also the statement that the electrical pressure is greatest at the left as the cause of the current towards the right, it is well to remark here that the electrical pressure at any particular point in a circuit is sometimes spoken of as its potential. If the potential at some other point in the circuit be different from the first, the current will flow from the higher towards the lower. The difference of potentials may be measured in volts, and expressed as E in Ohm's Law.

There is a very wide difference among different substances in their ability to transmit electricity. Some transmit it freely, and are called good conductors; others transmit it but slowly, and such are called poor conductors. All solids, and liquids possess some degree of conductivity; but some of them, such as glass, rubber, and wax, are so poor in conductivity as to be called non-conductors. The term non-conductor came into use before the refined methods now in use for measuring conductivity were known. It is now believed that the only non-conductor of electricity is the ether. If this be the case, then it appears that all the so-called electrical phenomena in the ether are to be looked upon rather as the results of electrified matter upon the ether, than the presence of electricity in the ether, just as radiations or ether waves are the results of actual vibrations of atoms and molecules. Conduction, then, is a general property of matter, and differs in degrees, that difference depending upon both the kind of element considered and its molecular combination. Thus, copper is an excellent conductor; but if copper be chemically combined with sulphur or with oxygen its conductivity is greatly impaired.

Conduction, too, implies contact, physical contact, as in the case of heat; hence solids and liquids may continuously conduct electricity, while gases can conduct no faster than their individual molecules can move in their free-path motions, and the rate of electrical loss is so slow from this source, that for telegraph lines of hundreds of miles in length it is neglected as being of no practical consequence. Neither is moist air much better, and for the same reason. In all cases where dampness appears to affect the working of electrical apparatus, the loss is due to the moisture deposited upon the surface of the apparatus, which thus forms a thin conductive coating. A Leyden jar may retain its charge for months if protected from a coating of moisture, which, of course, it could not do if either the air or the ether were conductors in any ordinary sense of the word.

The words conduction and conductivity represent the property possessed by matter to become electrified by mere contact with another body that is electrified; but the terms do not have a very high scientific importance now, for a much more convenient term is employed in place, the term resistance, which is the reciprocal of conductivity, that is, the greater the one the less the other proportionally. The substance having the highest degree of conductivity has the smallest degree of resistance. Resistance is measured in ohms, and is of two sorts; viz., specific and dimensional. Specific resistance is that resistance which depends altogether upon the nature of the particular element considered, and may be determined for any element by measuring the resistance of a cubic centimetre of it.

Tables of conductivity and of resistance of wires are common, and the following one gives the relative values of a few of the elements for comparison. The standard of conductivity being silver and reckoned as 100. The standard of resistance being a column of mercury 106 centimetres long and one millimetre square, which has a resistance of one ohm. The numbers given are the resistances in ohms and fractions of a wire 1 metre long (39.37 inches) and one millimetre ($\frac{1}{25.4}$ of an inch) in diameter.

SUBSTANCE.	CONDUCTIVITY.	RESISTANCE.
Silver,	100	.021
Copper,	99.9	.021
Gold,	80.	.027
Aluminium,	56.	.037
Zinc,	30.	.072
Platinum	18.	.116
Iron,	17.	.125
Lead,	8.5	.252
German Silver,	8.	.267
Hard Carbon,	1.	50.00
Graphite	0.01	Very variable.

The resistance of most liquids, and of such substances

as are used for insulating wires, is so very great that they are given in units called megohms, each a million ohms. The following represents the resistance of a few bodies in such terms, the volume being one cubic centimetre:—

SUBSTANCE.	RESISTANCE IN MEGOHMS.
Ice,	284.
Water at freezing-point,	150.
Mica,	84.
Gutta Percha,	450.
Hard Rubber,	28,000.
Paraffine,	34,000.
Glass,	3,000,000.
Air,	Infinite.

These must be read as so many millions of ohms. Thus, ice 284 millions. Thus can be seen within what wide limits this electrical property of matter ranges, and also its significance as a factor in Ohm's Law, and why some substances can be practically used as insulators when in reality they possess a certain degree of conductivity. Thus, glass is called an insulator. But if there were a difference of potential or pressure on opposite sides of a piece of glass one centimetre thick, equal to 3,000000 of millions of volts, there would be a current of one ampère passing through for

$$\frac{3,000000,000000}{3,000000,000000} = 1$$

In no artificial way can we produce such a voltage as that; but it is the opinion of some physicists that the voltage of lightning may rise as high as some thousands of millions. Under ordinary commercial voltages of only a few thousands, the current would be insignificant. Suppose it were 50,000 volts, then

$$\frac{50,000}{3,000000,000000} = \frac{5}{300,000000} = \frac{1}{60,000000}$$

of an ampère.

Dimensional resistance is of more practical importance, for by making a conductor larger its resistance becomes less. When the cross section of a wire is doubled, the resistance is reduced one-half. When the diameter of it is doubled, it is reduced to one-fourth,—a relation which may be stated as follows: The resistance of a conductor varies inversely as its cross section, or the square of its diameter if it be a wire; so by making a relatively poor conductor large enough, it may transfer as large a current as a much better specific conductor of smaller dimensions. In the table it is shown that the resistance of copper to that of iron is as .021 to .125, or that the latter is six times the former. If, then, the section of the iron wire be made six times larger, it will have the same degree of conductivity as the copper. This means that one pound of copper is worth nearly six pounds of iron for electrical conduction; and whether the one or the other should be employed in a given place depends chiefly upon the relative costs. It is a commercial rather than an electrical question. The resistance of all conductors varies with their length.

Temperature also affects the conductivity of nearly all bodies. Some have their conductivity increased by heat, as is the case with carbon; others have their conductivity increased by cold. Thus, the conductivity of copper at 100° below zero is increased nearly ten times.

An idea of the relative magnitude of the factor of resistance in common electrical work may be gained by knowing that a mile of ordinary electric arc-light wire generally has a resistance of about two ohms; telegraph and telephone wires five or six ohms, and often more, per mile. If there be a current of ten ampères going through a mile of wire that has a resistance of one ohm, then Ohm's Law enables one to determine what is the difference in pressure between the ends; for $\frac{E}{R} = C$ and $E = RC = 1 \times 10 = 10$ volts. So if any two of these factors be known, the other may be computed. The E gives the available electrical pressure; the R gives the conditions under which it can work, and the C gives their resultant, the available current, while the product of *EC* gives the activity, or rate at which energy is expended in the circuit, while if this product be divided by 746, the horse-power of the circuit will be given.

The further significance of Ohm's Law and its utility will be given farther on, when considering the relation of electrical energy to mechanical energy.

INDUCTION.

It has been pointed out that the term conduction signifies the transferrence of electricity from one body to another by contact,—contact in the sense that the molecules of solids and liquids are in contact when they cohere, and when their individual vibrations cannot take place without mutual interference. It is found that bodies become electrified by merely being in the presence of another body that is electrified, without material contact, and the more perfect the vacuum between the bodies the more freely does this phenomenon take place. As the electrified body that thus affects other bodies in its neighborhood does not lose any of its own electricity, does not share it with other bodies in any degree, and as the other bodies lose their electrification by simply being removed to a distance, and will recover it again by being brought back, it follows that the action is entirely distinct from the phenomenon of electrical conduction. A similar body electrified by conduction will retain its condition, and distance will make no difference. This kind of action is called *electrical induction*. To understand what changes take place, it will be needful to attend particularly to the factors present. Under the head of Electrical Origin of Electricity, it is pointed out that an electrical current may be induced in a circuit adjacent to another circuit in which a current is produced in any way; and here are similar conditions and similar phenomena. Imagine an electrified body freely suspended in the air. If a gold-leaf electroscope is brought within a few feet of it, its leaves will diverge; if brought nearer they will diverge still more; recession will cause them to collapse. This movement of the leaves can be produced indefinitely by changing the distance of the electrometer from the electrified body. It is important to note here that it requires the expenditure of energy to move the gold leaves, though the amount may be small. If it may be done for an indefinite number of times, then the energy spent may be indefinitely great; and that it is not directly derived from the electrified body itself is certain; for the latter loses by the process none of its electricity, and cannot lose it except by conduction. Evidently the body has in some way modified the physical condition of the space about it so that another body within that space is affected somewhat as it would be if touched by an electrified body. But the property belongs to the space itself, and cannot be extracted from it so long as the electrified body remains in place. This space about an electrified body within which other bodies assume an electrical condition is called an *electrical field*. It may extend to an indefinite distance from it, and its strength has been found to vary like gravity, being inversely as the square of the distance. This new physical condition into which the space has been brought by the electrified body is known to be the effect of the latter upon the ether, and is called its electrical *stress*. It is simply the reaction of the one upon the other, and indicates that the molecules stand in abnormal strained positions. A mechanical idea of what it is like may be got by pressing the hand upon the top of a table, and then producing a twisting strain tending to turn the table round, but without moving it. The whole table will be subject to a stress that will react upon the hand, a condition which will, of course, be retained by the table as long as such pressure is kept upon it. For the hand substitute an electrified mass of matter, and for the table the ether in any direction about it, and one will have a fair conception of the electrical field.

Especially so, if he will add to it that such twisting effect can be either right-handed or left-handed, and so produce those distinctions known as positive and negative, which run all through electrical phenomena.

A body brought into this distorted field of ether is acted upon by the latter tending to twist its molecules into new positions with reference to each other, which is precisely the condition that brought about the original stress, that is to say the electrical one, with this difference, that if the original one was right-handed the reaction of the ether would be left-handed, or exactly opposite that of the inducing body. This is simply because action and reaction are equal to each other and *opposite.*

One can now understand how it can be that bodies can be electrified by induction without loss of electrification by the inducing body. There are three steps in the process. 1st, The body electrified in any known manner. 2d, Its resultant stress in the ether. 3d, The reaction of the ether upon the second body, inductively electrifying it. Electricity has not been conducted by the ether, but a stress has been, and the ether stress has electrified the second body. By periodically electrifying and delectrifying a body, a series of stresses will be produced about it which will travel outwards as a succession of waves, the velocity of which is the same as that of light, 186,000 miles per second, and the wave length of which will depend upon the number of electrifications per second. Suppose a sphere, like a cannon-ball in free space, to be connected by wire, so that by pressing a Morse telegraph key in connection with any source of electricity it could be charged and discharged at will. If the key was closed regularly once a second, the wave produced would be 186,000 miles long. If it could be closed 186,000 times per second, the wave would be one mile long. And if it could be closed so often that the wave length should be but the one fifty-thousandth of an inch, there is every reason to believe that the eye would perceive the waves as light; not so much because the waves were produced by electrical means, as that the eye is capable of perceiving ether waves of that length, no matter how they may originate.

The analogy between heat and electrical phenomena in the ether is very close. The ether receives the energy from both sources and transforms it. The ether is not a conductor of either heat or electricity: it is neither heated nor electrified by them, but in each case is simply a medium for the distribution of such energy as gets into it according to its own laws, and quite independent of its source. When heat gives up its energy to ether it becomes ether waves or radiant energy, and is no longer heat: it has been transformed. When radiant energy falls upon other matter it is again transformed into heat. In like manner, when electricity gives up its energy to the ether, it becomes radiant energy also, and when this falls upon other matter it is again transformed into electricity. I have been thus particular to enlarge upon induction, and point out the factors present, in order to make it clear how entirely distinct the electrical condition in matter is from the electrical effect of it upon the ether. It is from a failure to keep these distinctions in mind that so many have been mystified by electrical phenomena, and so many different theories have been propounded as to its nature.

In all our experience electricity originates in matter, and whatever the particular character of the phenomenon *in matter*, it ought to have a different and distinct name from the effect of such phenomenon upon the ether. If such endowment of matter be called electricity, then it is not proper to use the same word for its stress, or wave effect, in the ether, and this for precisely the same reason as is allowed to hold good in heat phenomena. Formerly ether waves were called heat, afterwards heat waves, now radiant energy, for it is known that there is no heat in the ether.

EFFECTS OF AN ELECTRICAL CURRENT—1. MAGNETISM.

If a wire through which a current of electricity is passing be twisted into a loop or ring, it is found that the loop acts in all ways like a magnet. Its sides have polarity; and if it be so mounted as to be free to assume any direction, it will move so its sides face the north and south. If a piece of iron be placed in the ring, the magnetic effect will be greatly strengthened. Soft iron, however, loses its magnetic property as soon as the current stops. A piece of steel will retain some portion of the magnetic condition, and so is called a permanent magnet. A given current of electricity will make a much stronger magnet of a piece of soft iron than it will of a piece of steel, and this is explained by saying that the iron is more *permeable* to magnetism than steel is. Once in possession of a magnet, one may proceed to study its physical properties in many ways. That a magnet possesses poles; that it can attract and hold to itself iron, steel, nickel, cobalt, and affects other substances but slightly; that it is attractive to unlike poles of other magnets, and repulsive to similar poles,—and so on, are phenomena so widely known that they need not be described here. Only such phenomena will be considered as will be helpful to an understanding of the constitution of a magnet, and its relation to electricity and to the space about it.

The magnetism of a magnet seems to reside chiefly near its ends, for these will sustain bits of iron, but near the middle it will not; and when a small compass-needle is moved around a bar magnet, it points towards one end or the other, except when near the middle, where it sets itself parallel. When such a bar magnet has a sheet of paper laid upon it, and iron filings are sprinkled upon the paper, the filings are arranged in curious curved lines, starting from one pole and traceable to the other, and quite around the magnet on both sides. This arranging power of the

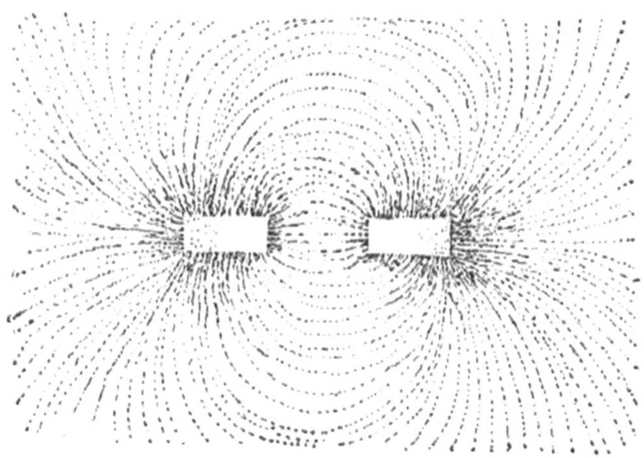

DIAG. 20.—MAGNETIC LINES.

magnet extends in every direction about it, as one can satisfy himself by trying the same experiment with the magnet turned on different sides. If one will compare the direction of these lines of filings with the positions of the needle, he will see that the needle assumes the same direction at any given place. Near the poles the lines all converge to it, and opposite the middle the lines are parallel with the magnet. If the magnet be of a U or horse-shoe form, the lines will be found still to extend from one pole to the other, some straight, some curved outwards, but always forming a curve such as to touch each pole of the magnet. While the filings are in the position described, let the paper be gently tapped with a pencil so as to jostle them slightly, and they will begin to close up in such a way as always to shorten themselves, and presently they will form a dense mass between the poles, adhering to the latter as a solid piece of iron would do.

Such phenomena show that the magnet in some way reacts upon the space about it, so that iron and other magnets there are affected, just as an electrified body affects the space about it, as has been described. This space about a magnet within which such effects are produced is called the *Magnetic field,* which may be said to be the stress in the ether produced by a magnet. Like the electric stress, it extends to an indefinite distance from the magnet, and travels with the velocity of light; so if a magnet was charged and discharged once a second, a wave motion would be set up: the wave length would be 186,000 miles long, and if it could be charged and discharged so fast that the waves were but the one fifty-thousandth of an inch in length, it is very probable they would be perceived as rays of light, and the magnet would be a luminous body. Such waves are called electro-magnetic waves. At present the shortest waves of this sort, that can be artificially produced, are several inches long, but it seems highly probable that before long some way will be discovered of making them of the required length for vision.

If a test-tube filled with iron filings be held near a delicate suspended magnetic needle it will be found to give no indication of polarity, one part will act just like any other part, and the magnet will be equally attracted. Bring the test-tube against the poles of a strong magnet for a few seconds, and then it will be shown that the filings have become magnetic, and now one end of the tube will attract one pole of the needle, while the other end will repel the same end. Shake up the filings well, and the polarity will be destroyed.

Stir up iron filings with melted wax, and pour into a paper mould, so as to form a stick the size of the finger, or larger. If this be tested for magnetism, it will be found without any; but magnetize it as if it were a piece of steel, and it will be found to retain it, becoming a permanent magnet. If a layer of iron be electrically deposited upon a brass wire in a magnetic field, the wire acts like a magnet. All these phenomena go to show that what is called polarity or magnetism is due to the *positions of the molecules,* rather than upon some sudden endowment which the molecules receive and may lose. Imagine every molecule of iron to be a magnet, having its poles or faces, then if in a mass of them, such as makes up a piece of iron or steel, all be made to face one way and keep such position, all will act in conjunction to give polarity to the mass. When some molecules face one way, and others adjacent to them face the opposite way, they will but neutralize each other, so the external evidence of magnetism will be destroyed. How atoms may be magnets and exhibit polarity may be imagined by considering the phenomena of vortex rings again. In the ring all the motion on one side is towards the middle of the ring inwards, on the other side all the motion is outwards, so the properties of the two sides are opposite. Each such ring must have its own *field,* which may extend to an indefinite distance from it, and may be represented roughly by the diagram in which the curved lines show the same features before described as belonging to a magnetic field. When two or more are facing the same way, and are in contact, these lines cannot re-enter the ring except by going round the second one; and when many are in line they must go round them all, in which case the direction of the lines will be precisely

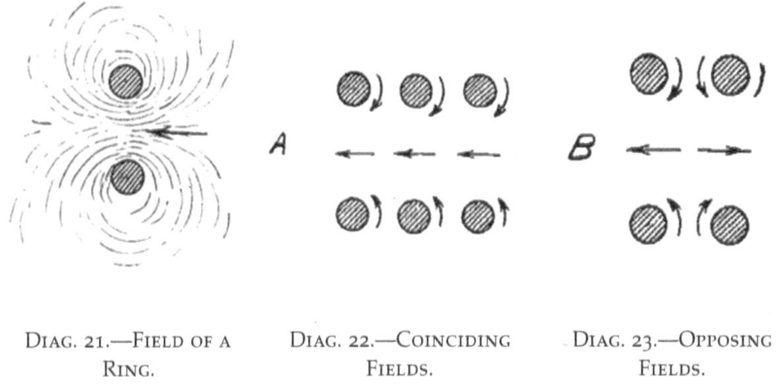

Diag. 21.—Field of a Ring.

Diag. 22.—Coinciding Fields.

Diag. 23.—Opposing Fields.

those observed about a straight bar magnet.

When they all face one way, as in diagram 22, the resultant will be at A, the sum of the outgoing movements, and at B, the sum of the ingoing ones, and polarity at A and B will be at a maximum. If they face in different ways, each will tend to cancel the other, and there will be no external field; as in diagram 23.

If two such atoms be brought face to face, each will be blowing against the other; their fields overlap, and the stress is increased between them, and they are crowded away from each other,—a phenomenon called repulsion. The opposite condition obtains when they face the same way and are near together, with the result that the stress is lessened between them, and they are pushed together by it; and this is called attraction.

There has been growing the conviction for a long time that the atoms of all substances are magnetic; but when they combine into molecular groups they are turned about so their magnetic fields neutralize each other, and thus it happens that most molecular compounds show no polarity. But every substance whatever is attracted by a magnet, and will move up to it if the magnet be a strong one. Brass, lead, stones, oats, corn, and wood will all be affected alike by a strong magnetic field, being pushed towards the magnet in the same way as iron, though not in the same degree. The pressure of iron against a magnet, due to the magnetic field, may be as great as a thousand pounds per square inch.

When a piece of iron is brought near to a magnet, and it becomes a magnet by induction instead of by contact, it is to be understood that its molecules are rotated into similar positions by the action of the magnetic field upon it, not that magnetism has gone from the magnet to the iron; and when it requires a pull, and therefore work, to move a piece of iron away from a magnet, it is against the ether the work is done.

It was stated at the outset that a loop of iron through which an electric current is passing is a magnet, and previous to that it was pointed out that an electric current in a wire has a field that extends indefinitely out from it. If such a wire be dipped in iron filings, they form rings round it, showing that the polarity is at right angles to the wire. Now, if the wire with the iron filings clinging about it be made into a loop, it will be seen at once how the polarity of the different segments is all in one direction inside the ring, and opposite to that on the outside the ring, and the structure will be a forcible reminder of a vortex ring. If several similar turns be taken in the wire, and they all be brought near together so as to form a helix, it will also be seen that these conspire together to set a boundary to the field on the inside, but allow indefinite expansion to it outside; so if one

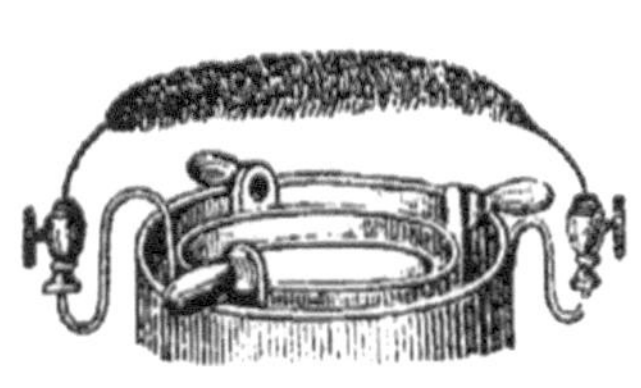

DIAG. 24—IRON FILINGS ABOUT ELECTRIC CURRENT.

should draw the lines for it as iron filings would be arranged by it, he will have the precise lines of a magnet, while the ring structure will be, on a large scale, just what was described on an atomic scale as constituting a vortex ring magnet; and the only thing lacking to complete the analogy is the conception of a rotary motion in the wire at right angles to its length.

It has been found that when a current has been started in a conductor, a torsional impulse is given to the latter in such a sense that if one looks along it in the direction of the current the twist is in the direction of the hands of a clock. So there is direct confirmatory experiment showing that the nature of the motion in an electric circuit is rotary in such a way that the whole circuit may be considered as a vortex ring; and as it is the matter of the conductor that is thus rotated, it follows that electrical current motion is rotary, as heat motion is vibratory.

Diag. 25.—Adjacent Turns.

Allusion has been made to the opinion now current that ether waves or light are electro-magnetic phenomena. How this can be may be understood by considering a magnet of any form, with its surrounding field. If the form of the magnet be changed, the shape of the field will be correspondingly changed; and as this extends out indefinitely into space, it follows that a succession of changes of form would set up waves through the whole of that space. Now, a magnet is an elastic body, and if it be struck it will vibrate and produce a sound. The vibration implies a change of form, and that in turn a set of waves radiated into space. As the field is an ether field, the waves will be ether waves. Now assume that atoms are themselves elastic magnets, each with a field indefinitely extended, and it follows that the vibrations produced by impact, or in any other manner, will set up corresponding waves in the ether, the wave length depending upon the vibratory rate of the atoms. Thus ordinary radiant energy, or light, would consist of undulations in a magnetic field.

Of course it will be perceived that vibrations of any electro-magnetic body, large or small, would induce similar waves, differing only in wave length, so there would be in the ether wave lengths of all dimensions, from the shortest possible to those millions of miles long. It is now an important physical problem how to produce such that shall be of the dimensions capable of affecting the eye.

Induction Coils.

One or more loops of iron, through which a current of electricity is flowing, is an electro magnet. When iron is placed in the loop, it condenses the magnetic field, and it may be made as much as thirty times stronger than it would be without the iron. When a magnetic field is produced inside a loop of wire, the reverse effect happens, and a current is generated in the opposite direction. Suppose a short rod of iron to have a single turn of wire at each end about it, one of them, as A, to be so connected to a source of electricity that a current through it may be produced by closing a key, the other one to be a closed circuit, as shown. If a current be established through A in one direction, a current will be induced in B, as indicated by the arrow. There will be in the loop of the A circuit a certain electro-motive force, *E*. A nearly equal electro-motive force will be induced in loop B. If there were two loops at B instead of one, the electro-motive force would be twice that in A, and for *n* turns it would be *n* times. The current in B will depend upon the resistance in its circuit; that is, it will

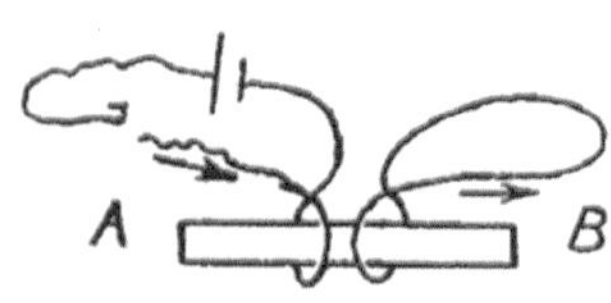

DIAG. 26.—ELECTRO-MAGNETIC INDUCTION.

be $\frac{E}{R} = C$, according to Ohm's Law. The size of the wire in B circuit will not make any difference in the value of E in it. That value will depend only upon the magnetism of the bar, and the magnetism in the bar will be measured by the product of the current into the number of turns of wire in the circuit A. And this product is called the *ampère turns*. The ampère turns will be nearly equal in the two circuits. This process of obtaining electrical currents in a second circuit by two transformations is of great use in the electrical industries, and the device is called an induction coil or transformer. The charging circuit is called the primary, and the discharging one the secondary. By making circuit A of a small number of turns of thick wire, so as to allow strong currents in it, and having circuit B consist of a great number of turns, the electro-motive force may be raised almost indefinitely. Suppose there be 100 turns in the *A* circuit, and a hundred thousand in the *B* circuit, then for every volt in the A circuit there may be nearly a thousand volts in the B circuit; and this is the construction in those instruments known as induction coils, with which so called jumping sparks are produced, and represent sometimes a million or more volts. On the other hand, it is sometimes desirable to change a high electro-motive force to a lower one;

and this may be done by reversing the connections and making the primary current go through a great number of turns, and taking the induced current from the smaller number of turns in the other circuit. Definite reduction in either way may be effected by making the ratio of the number of turns in the two circuits the reduction wanted. That is to say, if there are 100 volts in the primary circuit, and only ten are wanted, make the secondary of one-tenth the number of turns in the primary. If a thousand volts are wanted, make the secondary with ten times the number of turns in the primary. It should be remembered, also, that two turns of wire in B have twice the resistance of one turn, and the current induced will be reduced to one-half. If there be one hundred turns, it will be reduced to one-hundredth and so on. Hence, in the large induction coils for high electro-motive forces, the current is necessarily a small one, while in the transformers in which the reduction is to lower values of E than are in the primary, the current may be very great indeed. This is the case in Thompson's Welding Apparatus. The secondary has but a single turn of heavy copper, while the primary has many thousands, and the current in the secondary may be thousands of ampères. As the heating effect is proportional to the square of the current, it is plain that such large currents have enormous heating power.

All such devices require either intermittent or alternating currents to operate them, for there is no induced current in any circuit when the inducing magnetism is not changing. A constant magnetic field induces no electrical changes.

The Electro Magnet.

This is generally considered as consisting of a helix of insulated wire about a piece of soft iron, and may be either a straight bar, or crooked in any convenient form, its function being to produce a magnetic field when a current circulates in the wire, and to lose it when the current stops. This it does only partially, for all iron when once it has been magnetized becomes more or less permanently magnetic; hence there is only a difference in degree between an electro magnet and a permanent magnet. Until within a few years the electro magnet had its most extensive field of usefulness in telegraphy. It was combined with a piece of soft iron near its poles called its armature, which was so mounted that the magnetic field made it to move towards the magnet, and a retractile spring pulled it away when the field was absent. The movement of the armature was employed to receive signals. In some cases the movement recorded itself, and sometimes its prompt motion produced a sound, a succession of these being arranged into a telegraphic alphabet.

If one has a good idea of a magnetic field and its action upon a piece of iron in it, he will be able to understand all the various combinations of forms and functions of electro-magnetic devices, however much they may apparently be disguised. Thus, the magnetic telephone is an electro magnet with an armature of such size and flexibility as to be capable of much quicker movements than ordinary telegraph instrument armatures, the whole boxed so as to be convenient to hold to the ear. A common telegraph sounder acts in precisely the same way, though not so well, for the armature is too heavy, and one cannot concentrate its effects upon the ear on account of its form. An electric bell also produces its ring by having a hammer fixed to the armature, so as the latter moves in response to the electric field it strikes the bell.

An electric motor, in the largest sense, consists of a device for transforming electric into mechanical motions; and the relation sustained between an electro magnet, its field and an armature, is such as to do it directly. A telegraph sounder is thus a simple motor, for the armature moves visibly in response to the electric current. If a wire be wound about the armature, there is an induced current in it, as in an induction coil, and for the same reason; and the movements of the armature towards and away from the poles of the electro magnet, called sometimes the field magnet, give rise to currents in the armature coil. If a current from another source is sent through the armature coil, it gives polarity to the armature itself, and the reaction between it and the poles of the field magnet is still stronger, and the mechanical motions are still more energetic. The armature thus wound with wire is obviously an electro magnet itself, and when it is so mounted as to be capable of rotating between the poles of a fixed electro magnet, a continuous rotation may be kept up.

The current in the fixed magnet is steady, and therefore maintains a steady magnetic field. The current in the armature magnet is changed in direction by the motion of the armature itself, and is effected by a device called a commutator. The efficiency of such a motor may be as high as 90% or more. That is, for every horse-power of electrical energy turned into it, it will give back nine-tenths of a horse-power in actual work. The small space they occupy for the working capacity, when compared with a steam-engine for the same work, the small amount of attention they require, and their freedom from the dirt inseparable from an engine, commend the electric motor as a substitute for the engine in most places where power is wanted and an electric current can be had; for it is to be remembered that fifty horse-power can travel through a wire that can go through a gimlet-hole, while a steam-plant for the same work would require a large boiler and engine as well as a big chimney.

When the armature of a motor is made to turn by mechanical means, the shifting positions in the magnetic field develop electric currents in its coils. Such an armature cannot be turned as freely when the field magnet has a current in it as it can when it has not, and the energy spent in making it turn appears as a current. The device is called a dynamo, which may be said to be a machine for transforming mechanical motion into electrical motion. The steps are mechanical motion, magnetic field, electrical current; while in the motor they are simply the reverse,—electric current, magnetic field, mechanical motion.

The efficiency of a dynamo is very high indeed. It can transform 95% of the power applied to it into electrical power, and in this particular it is one of the most perfect machines in existence. There is absolutely no room for any important improvement in the dynamo as regards its efficiency. A good steam-engine may transform ten to fifteen per cent of the energy turned into it. A windmill may give fifty, a turbine water-wheel ninety, but when a dynamo gives ninety-five, it shows that the coming man has a margin of but five per cent for improvement in its efficiency.

Thus the magnetic field, which is simply the ether in an abnormal condition of stress, is the common agency between mechanical motions and electrical phenomena, and transfers energy one way or the other. All that determines whether it shall be one way or the other is simply which side has the excess of energy; for energy of a particular sort always goes from the body having more to one having less. Which side has the excess is determined solely by the mechanical conditions present.

Electric Lighting.

An electric current always heats the conductor through which it is passing. The amount of heat depends upon the strength of the current, and varies as the square of it. In a given circuit with a uniform current, the current has the same value, and therefore the same heating power, in every part of that circuit; but the temperature to which a body will be raised by a given current depends upon its own constitution, its size and electrical resistance. Connect together three wires of copper, iron, and platinum, each a foot long, and of the same diameter, and make them a part of the same circuit, so that the same current shall flow through them. If the current be increased gradually, the iron wire will grow appreciably warm, more current will make it hot; platinum wire will be only warm; while the copper wire will not have its temperature much changed. Still more current will make the iron red-hot, the platinum uncomfortably hot, and warm appreciably the copper; and more current will fuse the iron, perhaps make the platinum red-hot, but the copper may not yet be uncomfortably hot. This heating effect in a given wire is found to be proportional to its resistance: the iron wire having the greater resistance is most heated, and the copper having least, is least heated; hence to obtain a high temperature with a given current, a conductor must be chosen that has a relatively high resistance. Resistance, however, varies with the cross section inversely, so a small wire must be taken if the temperature of incandescence is to be reached with a small current; and a current that will raise half an inch of a wire to a white heat will raise a mile, or any other length of the same wire, to the same temperature; but the longer a wire is, the higher must be the electro-motive force in order to get the same current. For a given length of a wire the electrical energy spent in it will be found by multiplying its resistance by the square of the current, RC^2, which will give the products in watts, of which 746 equal a horsepower. Metals are liable to fuse and become useless, so that wires of carbon, made by heating organic fibres in the absence of air, as in making charcoal, are substituted for metals. They fuse only at extremely high temperatures; and being enclosed in a vacuum in bulbs of glass they cannot burn up as carbon does when exposed to the air when red-hot. This is the electric incandescent lamp. Most of them are so prepared that a current of about three-fourths of an ampère is required to properly light them, and this will be got when the difference of potentials between the lamp terminals is kept at a certain figure, so that lamps are specified by the number of volts they require, rather than the current; thus there are 50 volt lamps, 110 volt lamps, and so on. Now, such lamps take ordinarily about four watts for a candle, so a twenty candle-power lamp requires eighty watts, and that means $\frac{746}{80} = 9.3$ such lamps to the horse-power. Such lamps may last for a thousand or more hours. If a stronger current be used, they shine brighter, but their life is shortened. There is a process of slow disintegration going on in these lamps all the time. The surface molecules slowly evaporate under the vigorous vibratory movements present, and the carbon vapor thus formed sticks to the inside surface of the bulbs, giving them the familiar blackened appearance.

The Arc Light.

If an electro-motive force of forty or more volts be maintained in a circuit, and the circuit be broken at some place and the ends separated a small fraction of an inch, the current does not cease, and is maintained between the ends by what is termed an arc, where the temperature is so very great that almost all substances are reduced to vapor at once. All metals are fused and dissipated there. Carbon does not fuse there, but is slowly burnt up. The ends of the carbon reach a temperature higher than can be reached in any other known way, and the light they then give out is called the arc light. The rate of expenditure of energy in that small space where the brightness is, is generally some less than a horse-power. The current employed is about nine and a half ampères, and the electro-motive force about forty-five volts; hence $9.5 \times 45 = 427.5$ watts, and such a lamp may be equal to 800 candles, though they are generally rated as 2,000 candle-power.

By increasing the current the brightness increases, and there is no especial limit to the amount of light that may in this way be produced. With parabolic reflectors the light may be concentrated into a powerful beam. The inhabitants of Mars could see such a one, and it could be used for signalling between the two planets if the Martians had a similar one.

Seeing that the temperature to which a given conductor can be raised by a current is determinate, one can arrange for heating on any scale. There is no other reason than the relative cost of electric heating compared with the ordinary method with fuels, why it should not be in common use to-day. In most places the dynamo for the production of the current would be run by a steam-engine, requiring in its turn a furnace; and it is cheaper to use the fuel direct for heating, than to transform the energy so many times, each time with some loss. A common furnace is much more economical of energy than a steam-engine. But if ever electricity is obtained directly from combustion in an economical way, as there is some reason for thinking possible, electrical heaters will displace stoves and the common furnaces in the house. So the same current that lights the house will serve for cooking and warmth.

2. CHEMICAL EFFECTS.

When a current of electricity is passed through conducting liquids capable of being decomposed, such as acidulated water, and solutions containing more or less of the metallic elements, decomposition of the solution results, with the additional curious phenomenon that one of the elements of the decomposed compound appears at one terminal, and the other element at the other. Thus, if water be the liquid, hydrogen appears at one place and the oxygen at another. If the two terminals of an electric circuit were on opposite sides of the Atlantic Ocean, and a current were sent through the circuit, hydrogen would appear on one side and oxygen on the other. The oxygen is set free at that terminal at which the current reaches the liquid. The direction of the current being determined in the ordinary conventional way. Bring the wire carrying the current over and parallel to a suspended magnetic needle. If the current be going from south to north, the north pole will be deflected to the west. If the current be going from north to south, the south pole will be deflected to the west. Hence, if one looks along a wire in the direction of the current, oxygen will be given off at the next terminal if it dips in water. It may be convenient to know that when a battery is employed as a generator of electricity, hydrogen is set free at the terminal of the battery from which the current flows, and oxygen at the other end of that conductor.

The decomposition of water may be taken as a type of electro-chemical work; hence, when the mechanical conditions present where decomposition is going on are understood, they may be applied to any other case.

Under the head Chemical Origin of Electricity it was pointed out that the same factors which gave rise to the current also arranged the molecules of the liquid so that the oxygen sides of them all faced the same way, towards the zinc, which of course necessitates that the hydrogen sides should all face in the opposite direction. The other terminal of the battery tends to bring about a similar condition of things, so that between the terminals the molecules are all polarized or brought into an orderly arrangement. The direction of the electric current in such an arranged body of molecules in the liquid is from the zinc to the oxygen—oxygen, hydrogen, oxygen, hydrogen, and so on to the last molecule in the line, the hydrogen face of which is against the other terminal. So far this represents molecular arrangement, not molecular or atomic cohesion. There is good reason for thinking that dissociation of atoms in such molecules is going on all the time in some degree, on account of their incessant and vigorous vibratory motion. Such motion must tend to disrupt the atoms so that at any given instant there would be a relatively large number of atoms in the liquid already free and quite indifferent as to whether they recombine with the same or other atoms the next instant. If there be another agency present, like an electrical current, adding its energy tending to disruption, not only would a larger amount of dissociation take place, but when at one end of the line one element of the molecule, like oxygen, enters into a new combination which is more stable under the conditions present, the remaining hydrogen will combine with the oxygen of the adjacent molecule when that molecule is broken up, and so on along the whole line, leaving the hydrogen of the last liquid molecule to be freed against the other plate of the battery. This means that there is an exchange of partners among all the molecules of the liquid that take part in the current, else some of both oxygen and hydrogen would be set free elsewhere than at the terminals, which never happens.

Now, all molecules are combinations of atoms in definite proportions by weight, and it is therefore to be expected when such decompositions as the above take place the products will be found in the same proportions. It is the necessary outcome of the operation. So for every one part by weight of hydrogen set free, eight parts of oxygen will be liberated; and for a like reason twice the volume of hydrogen as of oxygen.

If a current of electricity be led through any liquid which it can decompose, and the material of the terminals be some substance that neither of the constituents of the molecule can combine with, both of the elements will be set free. Platinum is such an element; and if terminals be made of that, and dip into a tank of water, the current polarizes the molecules precisely as in the battery, and decomposition takes place in the same way,—oxygen being set free at the in-going terminal, and hydrogen at the out-going one. If the solution contains molecules of metallic salts of copper, nickel, iron, silver, gold, etc., the metallic side of the molecule faces in the direction of the current, the same as the hydrogen in the former case; and as a consequence, the metal is deposited upon the out-going terminal, whatever that may be, and the other constituent of the molecule is set free at the in-going terminal. For example, the sulphate of copper is a compound of copper and sulphuric acid. Where it is subject to decomposition by an electric current, the copper is deposited at the one terminal, and sulphuric acid at the other. If both the terminals be made of platinum, one will be covered with copper, and the other will be surrounded with the acid, and all the copper in the solution may be taken out. If the in-going terminal be itself of copper, the sulphuric acid set free will itself dissolve off the copper as fast as the acid is set free, and in this way the solution will be kept saturated. The metal may be deposited on any other metal. It is in this manner that electro-plating of all sorts is done. Each different metal requires different treatment from the others as to solution, electro-motive force, current per square inch section, and so on for the best results. To decompose water, as much as one and a half volts are necessary to initiate it, but copper salts require only a small fraction of one volt. The amount of decomposition in a given time, say a second or an hour, depends upon the current employed. A current of one ampère will in an hour decompose only about fifteen and four-tenths grains of water, liberating one and seven-tenths grains of hydrogen. The weight of other elements set free or deposited by an ampère per hour is determined by multiplying the weight of hydrogen set free by the electro-chemical equivalent of the element, and this is either equal to its atomic weight, or is one-half or one-third that. Thus, the electro-chemical equivalent of gold is $\frac{196.6}{3} = 65.5$, of silver $\frac{108}{2} = 54$, of copper $\frac{63}{2} = 31.5$, of nickel $\frac{57}{2} = 28.5$, and so on. So the amount of gold that will be deposited by an ampère in an hour is $1.7 \times 65 = 111.35$ grains; of silver $1.7 \times 54 = 91.8$ grains and so on. This shows a definite relationship between electricity and chemical reactions.

It is to be kept in mind that when substances combine there is always some transformation of energy, and heat is either absorbed or given out. When hydrogen and oxygen combine there is a large amount given out, 61,200 heat units for each pound of hydrogen. When, therefore, water is decomposed so as to set free one pound of hydrogen, the same amount of energy must be spent to do it. The electrical energy spent in a decomposing cell is, therefore, reducible to the heating effect, and may be calculated as such.

3. LUMINOUS EFFECTS.

When an electric current passes from one conductor to another through the air an electric arc is produced, and great heat and light are developed there. An arc is generally about an eighth of an inch long. By having a higher electro-motive force one may be made several inches long. The arc itself consists of the incandescent molecules of the air in its path mixed with some of the disintegrated particles of the carbon of the terminal. When an arc is formed in a partial vacuum the character of the phenomenon is very much changed. Instead of being concentrated into a narrow space, it spreads out into an oval form, the size of which depends upon the degree of exhaustion. The terminals may be separated to a much greater distance; the light becomes less intense, and shows as a kind of glowing gaseous globe, and this may extend to the walls of the glass vessel in which it is produced.

If the vacuum be made very perfect, no current can be got through it; for the ether is a perfect non-conductor. Even the spark from an induction coil that will jump several feet in the air will not jump a quarter of an inch in a vacuum. The jumping ability of an electric spark or current depends upon its electro-motive force. A thousand volts will jump but about the one-hundredth of an inch in common air, and ten thousand volts only about one-tenth of an inch. From such experiments it has been concluded that a flash of lightning probably has an electric pressure reckoned by hundreds of millions of volts, but there is some doubt about the calculation for such exceedingly high voltage. Glass tubes provided with platinum terminals hermetically sealed, and from which the air has been partially removed, when connected with the high voltage terminals of an induction coil exhibit phenomena that depend altogether upon the degree of exhaustion in the tube. If the air pressure be removed to about the one-hundredth of the normal pressure, the discharge appears as a broad band of purplish light between the terminals; if the reduction be to the thousandth, the light fills the tube. Still further reduced, the discharge appears broken up into striæ, or bright disks, their distance apart depending upon the degree of exhaustion, and they measure roughly the length of the free path of the gaseous molecules. If the exhaustion is carried to a very high degree, this free path may be made as long as the tube, or longer. This means that a molecule may move from one end of the tube to the other without coming into collision with another one.

When a molecule touches upon the electrified terminal, it is impelled from it with great velocity, quite like that exhibited in the radiometer, and probably for the same reason. It moves away from the terminal in a straight line in obedience to the first law of motion, and continues on till it strikes another molecule, or the surface of the tube, and it shines as it moves, on account of its vigorous internal vibrations; for each gas gives its characteristic spectral lines when thus made incandescent. Where they strike upon a thin piece of platinum they may make it red-hot by impact, and

DIAG. 27.—CROOKES'S TUBE. LONG FREE PATH.

where they strike upon the walls of the glass tube

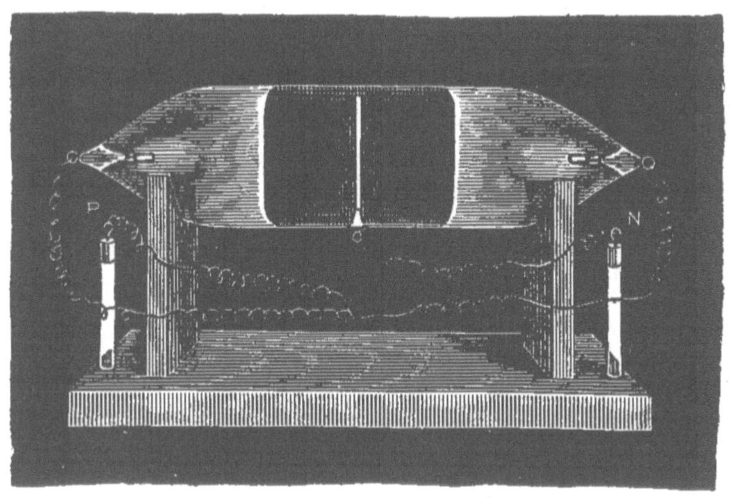

DIAG. 28.—CROOKES'S TUBE. PLATINUM MADE RED HOT BY IMPACT.

the latter is made luminous with a phosphorescent glow, and may be made red-hot, and so softened as to bring about a collapse of the tube. These tubes are known as Crookes's Tubes, and their phenomena are extremely interesting from the insight they give into the behavior of matter under all sorts of conditions. With a set of these tubes, the laws of motion, kinetic energy, sound, heat, light, electricity, and magnetism may be illustrated in a way unapproachable with any other simple and cheap apparatus. The long free path, and inability to turn a corner when projected from an electrified terminal, show the first law of motion and inertia. The impact of the molecules may make a wheel turn round,—an example of energy as good as a windmill. The intermittent beats upon the sides of the tube produce a sound, the pitch of which is the same as that of the vibrations of the induction coil. The heating of the tube and its contents shows the transformation of free-path motion into vibratory molecular motion. The luminousness of the gas, and the phosphorescence of the tube, show the transformation of the electrical energy into the vibratory molecular kind, at a rate capable of affecting the eye. The phosphoresence itself showing the conditions needed for producing it; the origination of the motions in the tube showing the relation of electricity to the other forms of motion developed; the deflection of the stream of electrified molecules by a magnet illustrating the effects of a magnetic field upon a current of electricity. The fact that such streams of molecules are projected from an electrified terminal solely by impact there, is shown by their returning to it when there is nothing in front of it to expend their energy upon, as a ball returns to the earth when thrown into the air, which is the case when but one terminal is connected with the induction coil; and, lastly, such a tube will be lighted up by being merely in the neighborhood of an induction coil, or rather in a varying electric field. They may be insulated and several feet away from such induction coil or a Holtz or other similar machine and yet be internally lighted every time a spark passes, which shows that the luminousness seen in the tubes is not necessarily due to any electrical current present, because in this case there can be no electrical current.

THE NATURE OF ELECTRICITY.

There have been many theories proposed to account for electrical phenomena, yet to-day there is no one that is generally held, even as a provisional one, among physicists. Some have even abandoned the hope of mankind ever being able to reach a consistent theory of it. The case has been the same in the history of heat, of light, and of magnetism; yet text-books of to-day do not hesitate to state what is the nature of each of these. Electrical phenomena have greater variety, and the apparent dual character oftentimes present has served to give a perplexing degree of complexity to them. The writer has thought that a summation of the principal factors present in electrical phenomena might be helpful to some in their endeavor to find some physical explanation without having to assume something *sui generis,* which has no other necessity for being except the very dubious one of accounting for a certain phenomenon. Caloric, light corpuscles, and vital force were such visionary creations; but further knowledge has enabled science to dispense with all of them, leaving nothing in their places but what was known to exist before; namely, matter, ether, and their motions. Such a steady course of reduction to these factors leaves one with the fair presumption that it will likely fare the same way with any other agencies that have been imagined to account for phenomena, though the latter may, for the time being, seem not reducible to simple mechanics.

There are certain *a priori* reasons for thinking that in electrical matters, as in all other physical agencies, only matter, ether, and motion are concerned. No one has ventured to identify ordinary matter and electricity, which cuts down the possibility to one of the remaining two.

If it be admitted that matter is not altered in quantity by any process to which it may be submitted, and also that the amount of ether and energy in the universe are constant, it follows that all the different phenomena exhibited by matter are due to the different kinds of motion it may have; for *motion is the only variable factor.* On such a premise one can fairly maintain that no matter how obscure and puzzling a phenomenon may be, its explanation lies altogether in its characteristic motions, and, when they are fully made out, there will be no more to learn about it. If so much be granted, one has got on a long way towards the final answer to the questions, What is the nature of heat? what is the nature of light? what is the nature of electricity? Two of these are settled, and no one thinks of asking as to their nature. The nature of heat was settled by Rumford and Davy, that it is a form of motion in matter. The nature of light was settled by Young and Fizeau, that it is a form of motion in the ether. What remained to be done was simply to discover the particular kind of motion in each case. Spectrum analysis and photography have since given us the particulars. Electricity is on precisely the same philosophical basis; and, in the absence of evidence of the existence of some other physical factors than matter, the ether, and motion, one would be entitled to the philosophical opinion that *electricity must be some form of motion*. What the particular form is may be a subject of investigation, but not the nature of it.

It is my purpose to show, *first*, that in every case where electricity is produced motion of some sort is antecedent to its production; *second*, that in every case the effect of electricity is to produce motion of some sort, and that itself is annihilated in doing it in precisely the same sense as motion of any other sort is annihilated when it is transformed.

1. *As to its origin.* When the face of the thermopile is heated and electricity is produced, we know that vibratory molecular motion is the condition for its appearance.

In a galvanic battery the molecular exchanges by which zinc is dissolved and oxidized, and hydrogen is set free, are well known, and also the heat equivalent of such re-actions; and they are measured in heat units, which in turn may be made the measure of the electricity developed.

When glass or wax or other substance becomes electrified by friction, the word itself expresses the condition necessary for producing it. Mechanical friction is the antecedent.

When a conductor is moved in a magnetic field and becomes electrified, the effect depends absolutely upon the motion. Stop that and all evidence of electricity disappears.

The same thing is true when electricity is developed by so-called induction in a field produced by a neighboring body that is electrified in any way. The continuous production of it implies continuous motion of one or the other body.

In dynamos of every variety of form the mechanical motion turned into them is the antecedent, and the energy of the engine spent in turning the dynamo has its full representation in the electric energy developed, and when there is no motion there is no electricity.

In the physiological development there are always chemical, thermal, and mechanical motions, which are spent to produce what electrical phenomena appear, whether in mankind or in animals.

In the air and in the earth there are changing temperatures, condensations, etc., which signify molecular motions.

Some crystals, like tourmaline, become electric by heating; some, like mica, become electric by splitting; and so on. Every one implies that some kind of motion has to be spent to develop the electrical condition, and in each case the particular kind of motion that has been spent to produce it has been *spent*; that is, it has been transformed in the same sense that the translatory motion of a bullet has been transformed into vibratory when it strikes the target. The electricity thus appears as the representative of the kind of motion that has been destroyed.

Some have imagined that electricity was a kind of dual matter, which was broken up by the various processes described, or that some substance was transferred from one place to another, so that there was more than the normal amount in one place and less in another. Even such conceptions do not get rid of the idea of motion being the chief characteristic, for the separations are the ideal embodiments of motion, and in this case the measure of it; so nothing whatever is gained, either in clearness or simplicity, by such invention.

2. *The effects of electricity* are to bring about mechanical motions of some sort.

The stress into which the ether is thrown by either an electrified or magnetized body is a change of position of adjacent parts with reference to each other, and the fact that this stress travels with the velocity of light shows that motion is the essence of it. The re-action of the stress in the ether upon other matter in it always results in the motion of the latter. If the whole body can move, it will do so, and mechanical motion is the immediate effect. If it cannot move as a whole, its molecules are twisted into new positions, so that motion, either molar or molecular, is the result.

As the electric current in a conductor always heats the latter in every part, one has but to reflect upon the character of heat motions to perceive that some kind of motion must be the antecedent of it. Consider a short portion of wire through which a current of electricity flows. It becomes warmer and now radiates faster into space. It is losing motion by imparting it to the ether. Trace back the ancestry of the ether motion, and it appears as vibratory motions of the molecules of the conductor, thence as electrical current, thence as armature rotations of a dynamo, thence to the engine movements, thence to the furnace and the chemical re-actions going on there. There is no question as to the nature of the factors in all of these but one. Call the chemical re-actions *A*, the engine *B*, the dynamo *C*, the electricity *D*, the heat *E*, and the ether waves *F*. With the exception of *D*, each one is known to represent a certain kind of motion, molar or molecular, and all in a consecutive series. Is it not difficult to conceive that the step *D* can be anything different in character from the rest of the series, and, whether understood or not, must represent some phase of motion? To think otherwise is to think that motion can have some other antecedent than motion. Whoever sets himself in earnest to this problem will see there is but one answer to it.

So heat effects, light effects, chemical effects, as well as the direct mechanical ones shown in Crookes's Tubes, or otherwise, will lead to precisely the same conclusion that *electricity represents an intermediate molecular kind of motion*, having definite motions for its antecedent, and definite motions for its consequent, and so must itself be some peculiar form of motion, differing from the others as they differ among themselves, and nothing beyond that. It may also be remarked that every form of motion which is capable, under definite mechanical conditions, of developing electricity, electricity is itself capable of producing. The processes are all reversible. If heat will produce electricity, electricity will produce heat. If chemical re-actions produce electricity, electricity will produce chemical re-actions, and so on of all the rest; so if they be reducible to motions, so must electricity.

Such considerations make logically certain what the nature of electricity is; but they do not indicate what the character of the motions is that gives it identity, and distinguishes it so radically from other well-known kinds of motion. In the chapter on "Motion" it is pointed out that there are three fundamental kinds of motions,—translatory, vibratory, and rotary,—and that with these all the various complicated motions of mechanical processes may be produced. It is also pointed out that for convenience we call those motions mechanical that are on a scale of visible magnitude, but such as cannot be seen are called molecular and atomic. It is plain, in this case, that the motions must be on a molecular scale, for no motions are directly perceived in electrical phenomena any more than in heat phenomena; so there remains for consideration what evidence there is for the motion being molecular and therefore of matter, or of the ether.

It appears that when certain kinds of work, such as friction, are spent upon a mass of ordinary matter, electricity is developed, and we say the body is electrified. The body in this condition at once re-acts upon the ether about it; and it has happened that some persons have given most attention to this effect of the electrified body, and the phenomena that may result from it, and have called *it* electricity; while others have given more attention to the condition of the matter that induced the ether stress, and they have called *that* electricity; while the greater number have hopelessly confused the two, calling both by the same name, just as formerly heat and ether waves were both called heat. It is plain that a physical condition of things in matter requiring a name ought not to be designated by the same term as that physical condition in the ether which is the result of the first. One is, therefore, justified by the logical necessity of making a distinction, in adopting the name electricity as applicable to one and not the other, and also in calling the phenomenon in matter by that name and denying its applicability to any effect of it wherever it is plain there has been a transformation. Thus it would be as illogical to call ether waves set up by an electrified body electrical waves, as it would be to call the swinging of a pendulum that was actuated by electrical attractions electrical vibrations.

We are, therefore, now reduced to the sole consideration as to the character of those molecular motions which differentiate electricity from heat and free-path motion; and here the apparent dual character, which has been so puzzling, helps at once to an understanding of it.

For many years it has been merely a matter of convention that a current of electricity is said to move in a certain direction in a wire. It has often been noticed that there is an apparent current in both directions from any electrical source; and one has been called a positive, the other a negative, one; yet the current, reckoned either way from its source, is always the same at a given point, and has not unfrequently been considered as made up of two currents moving in opposite directions.

If one will take a limp rope a few feet long and tie its ends together so as to form a ring, and, holding it in his two hands, will begin to twist it in one direction, he will see the twist start in opposite directions at his hands, and each one can be traced quite round the ring, neither interfering with the other; yet one is a right-handed twist, the other a left-handed one; and one might call one a positive and the other a negative current. There will be as much twist in one part of the rope as in any other, and the rate of rotation at the hands will be the measure of the amount of motion, and, consequently, of the energy that is in the circuit. For a rope substitute a wire, and for the hands a battery or a dynamo, and the analogy is complete, except that no rotation is seen in the wire as a whole; so, if there be rotations, they must be of molecules and not of the mass. Molecular motions must, of course, be inferential. It is so for heat. The waves called ether waves imply vibrations of matter; and, if there be any known rotary motions in the ether, they would imply molecular rotations for the same reason. It is conceded that in every electro-magnetic field the ether is in a rotary motion, and in numerous books it is pictured as a whirl both about a magnet and a wire carrying an electric current. The rotation of an electric arc in a magnetic field shows it, and the twist given to a polarized ray of light in passing through it also shows it; and it has been so interpreted for years. The twist given to a conductor through which a current is flowing, which has been before alluded to, also gives direct evidence of the same condition; so the phenomena confirm the conjecture that the phenomenon in matter which is called *electricity is a phenomenon of rotating molecules,* in the same sense as the phenomenon called heat is a phenomenon of vibrating molecules.

If the atoms in molecules, and the molecules themselves, were absolutely fixed in position so as to have no individual freedom of motion, there could be neither vibration nor rotation; but the vibrations tend continually to separate them, and hence between impacts there is freedom for rotary slip, if there be any tendency to do so. In an electro-magnetic field the ether stress re-acts upon molecules in it so as to rotate them upon some axis tending to set them in certain position with reference to it. This action will be stronger upon an atom or molecule immediately adjacent to an electrified molecule than to one more distant, and one may therefore infer that the process called conduction, where heat is the immediate effect of an electric current, is really an induction effect, and depends directly upon the ether rather than upon the direct mechanical effect of one molecule upon another; for such mechanical action would make the rotation of adjacent molecules to be opposite in direction, whereas in an electric current all are in one direction. There is, therefore, impact and slip, impact and slip; each impact knocking the molecule out of the position the induction had set it in, and each arrest of the slip resulting in increasing the amplitude of vibration, and hence raising the temperature of the conductor. Hence, the explanation of the transformation of electrical energy into heat energy. An electric current is, therefore, not a simple phenomenon, but is considerably complicated, involving motions of both molecules and the ether; the molecular motion depending directly upon the re-action of the ether stress produced by an adjacent molecule rather than upon mechanical contact. The electrical condition called static being itself a compound of abnormal molecular position and stressed ether, is the condition which, while being propagated in a conductor, constitutes an electric current, propagated in the ether, constitutes an ether wave.

CHAPTER IX

Chemism

THE atomic theory of matter was held in some form by ancient philosophers, but the reasons they assigned for their opinion were not such reasons as have led men of the present day to adopt that theory to the exclusion of all others. Modern chemical analysis enables one to reduce compound substances to their elementary forms, and out of those to build up numerous other substances with entirely different qualities. Each such elementary form can be isolated, its properties can be studied, and by compounding them one can at will produce thousands of substances, each with its own distinctive qualities. Some of the more thoughtful men of all ages have pondered upon the fundamental questions of physical science, and they have guessed how it might be: some guessed this way, some guessed that, and none of them gave a sufficient reason. It would be very remarkable if, among a multitude of guessers, some did not guess nearer right than others; but such lucky guessing hardly entitles one to the honor of being the founder of a philosophy that had to wait for later men and entirely different methods to substantiate it. And this is the real state of the case in nearly all departments of knowledge. Ask any chemist to-day why he holds the atomic theory of matter, and he will reply that he can isolate the elements, and by no process yet discovered can they be more finely divided; that he can measure their individual magnitude and weigh them, prove their existence in the sun and stars; so that the weight of evidence is exceedingly great. He will never think of assigning any such reasons as the early philosophers gave for their teaching. Many of the properties of bodies of visible magnitude depend upon the number and arrangement of the molecules that compose them, but the properties of atoms are fundamental and not subject to change. All substances are identified by means of their properties, and the chemical properties of atoms are among the most important. Not only do atoms combine together in groups called molecules, consisting of two or more atoms, but they combine in definite proportions by weight, and only so; and these proportions are called the atomic weights of the elements, and are known for all of them; so that molecules are compounds of definite constituents, definite weight, and possessing definite properties. For instance, water is made up of hydrogen and oxygen, two parts by weight of hydrogen and sixteen of oxygen; and as to its properties, such as density, specific gravity, conditions at different temperatures, etc., all are familiar with. Most of these properties of bodies are called physical, but by chemical properties is meant the ability of atoms to enter into definite combinations with other atoms, to form new compounds and develop new properties. The chemist is concerned with such atomic exchanges, called re-actions, and notes the conditions under which they take place, and some of the new qualities that appear, such as its physical condition, as to being a solid, a liquid, or a gas at certain temperatures, its crystalline form, if it has any, its behavior with polarized light, and so on.

Underneath all chemical re-actions there lies the question as to why atoms combine at all. At first it was explained as due to an attractive force,—chemical attraction, possessed by all atoms, but in different degrees by different elements. When it became known that this acted in definite selective ways, it was called chemical affinity, but was still supposed to be a peculiar force unrelated to other forces supposed to exist, such as heat, light, electricity, and so on. In the progress of knowledge, it became apparent that these latter phenomena were so directly related to each other that they were capable of being transformed one into the other, and then the expression "correlation of forces" began to be used. A further analysis showed them to differ from each other chiefly in the character of the motion involved in the phenomena; and so forces, as such, have been banished from physical science, leaving not even a single primal force; for as each one can be changed at will into any of the others there is simply a closed chain of phenomena, no one of which can be called an elementary one more than any other.

Chemical phenomena have been found to be a part of the same grand division, and the term "chemical affinity" has itself been in a measure supplanted by the term "chemism," which is now used to signify the quality possessed by atoms to enter into definite combinations; and its explanation is to be found by noting the factors present when atomic and molecular exchanges take place, and these have been found to be all physical without exception. There is a large field known as chemical physics with which one needs to be acquainted in order to understand simple chemical operations; namely, the effects of heat, light, and electricity in bringing about chemical changes.

When hydrogen combines with oxygen to form water the process is called a chemical one; but, as has been pointed out in the subject "Heat," there is a definite amount of heat given out by the combination of a definite amount of the elements; and in like manner the dissociation of the elements in water requires the expenditure of energy proportionate to the amount decomposed. This too is called a chemical process, but the conditions for doing either are purely physical, depending absolutely upon heat. The elements cannot combine when heat cannot be given out, and cannot be separated except by an equal expenditure. What is true for this example is true in degree for all other chemical re-actions; physical energy is involved in every change and is the condition for the change. The first law of thermo-dynamics states the quantitative relation between heat and mechanical work; viz., that it is measurable in foot pounds, and is equal to 772 foot pounds per pound degree, and this is called a heat unit. Now, the chemical combination of a pound of hydrogen with oxygen gives 61,000 heat units, and is therefore at once measureable in foot pounds, showing a direct relation between chemical re-actions and heat or work.

It has also been discovered by experiment that in the absence of heat chemical re-actions cannot go on, and this has led chemists to the conclusion that at absolute zero chemism does not exist. There is not only no selective action, but no cohesion among atoms, and all molecules would fall to pieces—that is, to atoms, quite dissociated—at absolute zero. Instead of requiring 61,000 heat units to dissociate a pound of hydrogen from water, it would not require any, for if the atoms do not cohere, no work would need to be done in order to separate them.[1]

From this, then, it appears that chemism is determined by heat, and does not exist in the absence of temperature. When it is developed it manifests itself in selective ways, and in the formation of definite compounds; and it therefore is a proper subject of inquiry as to how the temperature of atoms can give such selective qualities to them. This requires a reconsideration of the distinctive quality of heat itself. It has been pointed out that this consists in the internal vibratory motions of atoms and molecules, as distinguished from translatory and rotary motions; that the evidence for this comes, first, from the fact that a body of any size possessing any degree of heat—that is, having a temperature above absolute zero—is constantly exchanging its energy with the ether, and that the rate of the exchange depends upon the temperature; and, second, that translatory motions of bodies in ether do not require the expenditure of energy, or, in other words, that for such motions the ether is frictionless. This is the same as saying

---

[1]See Appendix, p. 477.

that, where the heat of a body is lost by radiation, it is the internal vibratory motion alone that is lost, not its translatory velocity. Consider a body of any magnitude whatever, having any temperature whatever, and moving at any assignable velocity in space. After an interval it will have lost some of its temperature by radiation, and, if it moves long enough, it might lose it all, reaching absolute zero; but its translatory velocity will not therefore be reduced in any degree. Hence, in considering the heat in a body, independent of any other motions it may have, one has only to do with its internal vibratory movements, and that the temperature of a body, say an atom, is measured by the amplitude of its vibration, and is proportional to the square of that amplitude.

If, therefore, chemism is directly related to heat, one must attend to what must be going on in an atom, not groups of them.

To say that an atom vibrates is to say that it is changing its form, and to explain how changing its form can result in such selective properties as atoms exhibit is to explain chemism by the mechanics of the motion involved. Whether atoms have one form or another will make no difference in this argument, which is that the result is due to change of form, whatever that may be; but, for making the subject mechanically clear, some form may be adopted, and one can do no better than to choose that form which now has most probability in its favor judged by other phenomena; that is, the vortex-ring, which has been treated under the head of "The Ether."

When such a body vibrates in its simple way it elongates alternately on two axes at right angles to each other; that is, the change in form is from a circle to an ellipse, so as to assume first a horizontal, then a vertical elliptical form, as shown in the cut. Such changes are due to the elasticity of the ring, and are brought about in such an atom by impact, by friction, and by absorption of ether waves. Whether produced in one way or another, they represent absorbed energy and exhibit it as heat, the temperature of a given one depending upon the amplitude given to it by a definite amount of energy however applied.

DIAG. 29.

Such changing forms imply nodes and loops in the vibrating body, positions of minimum and maximum motions; and when the vibratory rate is the fundamental one,—that is, the lowest rate the body can have,—there will be four of each, the nodes being the positions of minimum change of form. Such nodes may be seen in vibrating bodies of all sorts,—strings, bells, rods, pipes, and rings. The size of a body makes no difference in this characteristic, and it therefore may be affirmed of atoms as well as of any other magnitudes.

Let it be admitted that vibrating atoms can cohere for any reason, it will be seen that an atom such as represented could only have other atoms attached to it, and be in a stable condition, when they were at the nodes; and in this case four might be so attached and no more, if they were approximately of the same size. Such places in atoms might be called bonds: they would be definite in number, position, and strength. If the other attached atoms were themselves vibrating, they would each have their own nodes; and if they were free to turn into any position, one might be sure that the nodes of each would be in contact, and that the loops of the vibratory motions would be where space to move in without interruption was free. Such a combination of atoms might be called a molecule. It would consist of a definite number of atoms, each with its own atomic weight; and if the strength of the cohesion depended upon the vibratory motion, it

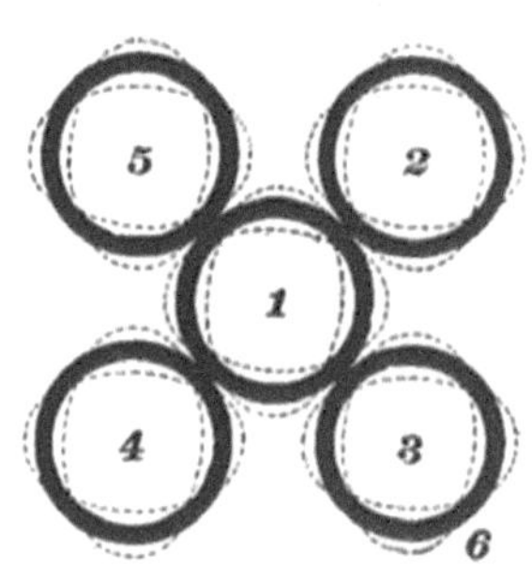

DIAG. 30.

Geometrical Forms of Snow Flakes.

is easily seen that when there was quiescence in that there would be disruption or dissociation. Moreover,

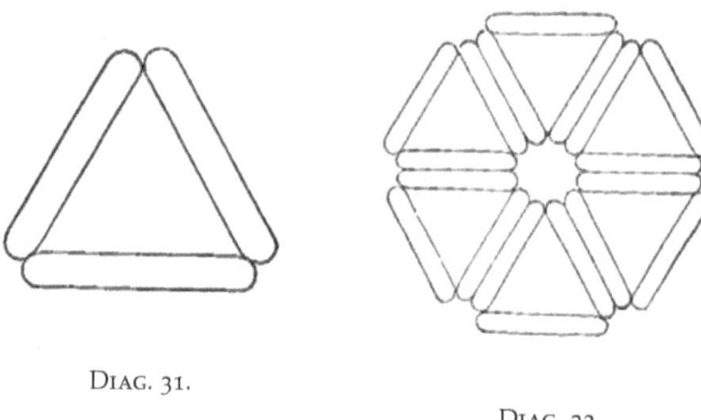

DIAG. 31.

DIAG. 32.

when there was such a nodal bond it would be like a hinge, and two thus united could swing upon it; while if three were thus united and two were to swing upwards, they would meet at a node on each and stick together for the same reason the other nodes did, thus forming a symmetrical and stable figure against which other similar ones could be built up, node against node indefinitely. A hexagonal figure would result. If four were attached to the primary nodes, and each was to swing up ninety degrees, there would be formed a sort of cubical box without a lid; but at the top will be presented four open nodes, upon which the four nodes of any other similar one might be placed: and thus could a cubical structure be built by addition of similar forms indefinitely. Such symmetrical forms are called crystals.

Of course all this presupposes that there is some good mechanical reason for atomic cohesion, that is in some way dependent upon temperature; and to make this clear it is needful, first, to call to mind some phenomena of a similar sort on a larger scale.

It is well known that if a light body be brought near a vibrating tuning-fork, the latter acts as if it attracted it, for the light body will move towards the fork. The same thing is true of other vibrating bodies, and the explanation is that the vibratory motion reduces the pressure about the body. Thus, suppose the hand to move to and fro; as it moves forward the air in front of it is somewhat condensed, while that behind it is partially rarefied; when the hand returns the same thing happens. The air follows up the hand because the pressure is reduced next the hand, and if the hand could swing back and forth, faster than the air could return to it, there would be formed a perfect air vacuum; and that means that the pressure would be nothing at the hand and fifteen pounds per square inch at a distance from it. Hence any body placed near the hand would be subject to a pressure greater on its remote side than on the side adjacent to the hand, and would be pushed by it towards the hand. This would be a phenomenon similar to attraction, the movement towards the vibrating body being due directly to the pressure of the medium, while the difference in the medium would itself be directly due to the vibratory movement. The amount of such difference in pressure is evidently determined by the degree of vibration. Now, if one can imagine a similar condition of things about an atom vibrating in the ether, he can understand how its vibratory movements might reduce the ether pressure adjacent to it in a way proportional to the movement, and also how at the nodes such effect would be at a minimum, and at the loops at a maximum, so there would be produced what is called a field. As the condition that produced it was one of mechanical motion, one might call the field a mechanical field, for mechanical effects of translatory motion result from it.

When such an effect takes place among atoms one might distinguish it as a *chemical* field, for it would bring about mutual cohesion among atoms, and the nodes would determine the positions of stable combinations; and a molecule so built up would require an amount of energy spent upon it to break it up equivalent to the energy spent, to produce the field, or, in other words, equivalent to the heat in the atom.

It is here to be noted that when atoms combine in

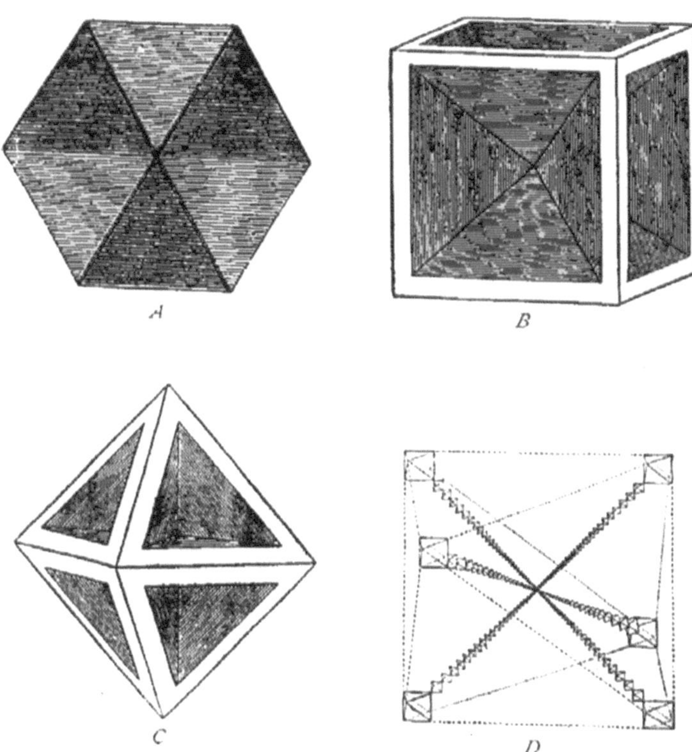

CRYSTALLINE FORMS.

The above figures illustrate very clearly the molecular arrangement in crystals of various kinds. *A* represents a cross section of Brazilian Topaz, as shown in polarized light. *B* is a hollow faced cube of salt, and *C* a similar hollow faced octahedron of copper sulphide. They show that the cohesive strength is greater on the edges than elsewhere. Some crystals, when being dissolved, leave a complete skeleton of themselves the last to disappear. *D* is a skeleton crystal of silver from Scotland, where the structure consists of a series of minute octahedral crystals adhering to each other in such directions as would build up a single large octahedral crystal if filled out.

this way each one retains abundant space for its heat movements, so its temperature may be varied within considerable limits without interfering with molecular stability. And, if the vibratory movements continue, then each molecule will have its own field, which will be the resultant of all the fields of the atoms that are combined thus to make the molecule. The field of a molecule will then have a form which will depend absolutely upon the number and arrangement of the constituent atoms, and will extend to some distance in space beyond the geometric boundary of the molecule itself.

The presence of such a chemical field must affect other chemical fields in the neighboring space where the fields overlap, hindering or facilitating the exchange of atoms in other molecules, because lessening the pressure holding them together. There are many examples of this kind of action known. It is called catalysis, which signifies the action of a given substance in bringing about chemical reactions without itself being changed. For example, the binoxide of manganese, when mixed with the chlorate of potash, greatly facilitates its decomposition by heat, though the binoxide is itself not decomposed. Pure zinc is dissolved with difficulty by sulphuric acid; but a little mercury or iron, or other so-called impurity, enables it to be dissolved freely.

Hydrogen and oxygen gases will not combine when simply mixed; but a little spongy platinum placed in the mixture will at once bring about the combination, but will itself suffer no chemical change. These gases will also slowly combine in the presence of mercury when kept at the temperature of 305°. In glass vessels without the mercury no combination at that temperature occurs, but on raising the temperature to 448° it combines very slowly. In smelting operations a flux has a similar function, and in some cases the boundary line of such action can be observed. Some re-actions take place at a different rate near the sides of the vessel that contains the solution than away from it, and some mixtures of substances in solution will separate from each other except within a short distance from the surface. Such phenomena show that the mere presence of some substances is sufficient to profoundly affect chemical re-actions. The chemical field of substances gives a consistent explanation of catalysis. There is another class of phenomena well known, but hitherto without any rational explanation; viz., some supersaturated solutions seem unable to initiate the process of crystallization, but the smallest crystal of the substance starts it, and the whole body is solidified in a few seconds. Here it is evident that the crystal, taken as a nucleus, had a field that compelled other and similar molecular groups to arrange themselves in similar order. This is a phenomenon of such importance as to warrant some attention here. When two tuning-forks having the same pitch are separate from each other a distance of several feet, and one of them be made to produce a sound, the other one will be made to sound likewise by the action of the sound waves in the air upon it. The effect is called sympathetic vibration. Other forks having different rates of vibration will not be similarly affected, so the vibrations in the air select out the particular fork having the same rate as the one vibrating, and cause it to enter into a similar state of vibration. So it appears with a magnet. Any magnetic bodies in its field become magnetized there; that is, they are brought into the same physical state as the body that incited the field. Such physical fields, then, are capable of compelling bodies in them to assume the same state of motion or similar position, or both, as the body that produced the field, provided the substance itself be constituted molecularly like the first,—and this simply by being in proximity, not by contact. It is a kind of induction, common through the whole domain of physics. In the organic world of living things the phenomenon of growth is manifested by what are called cells, which are symmetrical groups of molecules, as crystals are, only much more complex.

Growth consists in the formation of similar cells out of suitable molecular constituents in the neighborhood. Each different part of a plant or animal has a different cell structure. If, therefore, it be conceded that each cell has a field, which is the resultant of all the elements that make it up, it will be seen how such field must act upon other matter within it, compelling it to assume a form similar to the cell that produces the field; that is, to form a similar cell adjacent to itself. Such formation is called growth; but the similarity in form and function, when appearing among plants or animals, has been considered as due to heredity, a term that has a definite enough meaning, but which has not been supposed to be due to mechanical necessity but to some super-physical agency not amenable to purely physical laws and conditions. It is possible to pursue this much further and to show that cell structure itself may be modified by molecular fields, and how stability of form and function are possible with some and not with others,—how what in natural history is called variability, reversion, and other phenomena of the sort, are explicable as due to the same factors that organize atoms into molecules, and molecules into crystals. Every one interested in the fundamental questions of chemistry will be able to follow out in many ways the mechanical conceptions here introduced, and compare what he knows of chemical re-actions with them. It will be especially helpful for one to draw upon paper such ideal atomic rings with their edges touching, and marking where the nodes must be. Such diagrams as the one on p. 294, fig. 30, thus drawn, cut out, and the parts bent up until they touch each other, will probably surprise one at first to find how the nodes will be brought adjacent to each other and therefore into a stable position.

So far it has been assumed that there will be in the ether about a vibrating atom an effect comparable with the effect produced in air about a tuning-fork or other vibrating body that is producing sound waves. One might be satisfied that there was such an action, even though he were not able to explain it, provided there were good reason for the assumption. The case is the same as for a magnetic field within which magnetic phenomena take place, though a magnetic field cannot be isolated. It is the same for the existence of the ether itself: it is inferential, but from a large body of phenomena of different sorts, all corroborating the hypothesis; so one is satisfied. When a magnet acts upon a piece of iron not in contact with itself, we explain the action by the magnetic field; and, if a body acts chemically upon other bodies not in immediate contact, controlling their motions and positions, as is the case, the same kind of an assumption is to be entertained. If a reasonable explanation for the existence of the field can be offered, all the better, though no one holds more lightly upon a magnetic field because he cannot explain it. In the chapter on magnetism it is remarked that there is good reason to think that atoms of all kinds are magnets. If that be the case, then every atom has a field of its own, wherever it may be; and it would seem likely that this magnetic field of atoms was the underlying factor in the so-called chemical field; and it is therefore well to analyze the phenomena, having that magnetic field in mind.

A single magnet of any form will have its field under all conditions, and the *shape of the field will be determined by the form of the magnet.* If the magnet were of sufficient size, there would be no difficulty in locating it by its field, even though the magnet itself could not be seen. A number of magnets arranged promiscuously would so neutralize each other's fields as to have no residual field, and in order to detect the existence of magnetism it would be needful to get very close to an individual magnet. When a steel magnet is dissolved in an acid all evidence of the existence of magnetism disappears, for the iron molecules are now separated from each other and are scattered promiscuously through the solution. Any disturbance whatever that disarranges the magnetic arrangement of molecules destroys the evidence of the magnetic field, except at very short distances. When a piece of iron is heated to redness it cannot be made magnetic in the ordinary sense; for the vigor of the vibratory movement continually knocks the molecules into new positions, and therefore changes their resultant fields, leaving but a neutral effect upon outside bodies.

As chemical re-actions take place in liquids or gases, and only exceedingly slow in solids, it follows that in them one has to deal with molecules in all positions,—that is, an entirely disordered arrangement, and such as would exhibit no evidence of magnetic field, even though every atom was itself a strong magnet; and this condition of neutrality would be constant so long as the temperature kept up so much mechanical disturbance as to prevent any systematic arrangement. Yet it is to be borne in mind that the magnetic field of no one has been destroyed: it is as strong, as far reaching, as ever; but it is masked by overlying fields,—that is all. Let any one of them suffer any change at all, and the effect of it would be felt throughout the whole space the field would occupy if there were no other one in its neighborhood.

Now, when the form of a magnet is changed, it changes the form of the magnetic field—that is, the distribution of the stress that constitutes the field; and, when an atomic magnet vibrates, it is changing its form; and as a result its field is changing at the same rate. A multitude of such independent magnets, all changing their forms and fields, would be sending out waves into the ether; but they would be caused by and measured by their heat motions, not by their magnetic condition simply; and the effects of these waves at a distance from their source would be practical uniformity unless the waves were very long. For such short ones as are produced by atomic and molecular vibrations there could be no ordinary indications of a magnetic field such as are exhibited in the movements of bodies of visible magnitude. Long waves of precisely the same sort caused by motions of a slower rate might make magnetic needles move. Thus, magnetic needles upon the earth have been observed to move at the same instant that solar disturbances have been witnessed through a telescope, which indicates that the waves were long ones, giving a magnet time to move one way before it was impelled to move in some other way.

This condition of practical neutrality on account of the rapidity of the change at a distance from the magnetic body would not hold true in close proximity to the body itself; for the changes in the field will not only be actually greater there, but the fact that there are nodes and loops necessitates changes in the stress at the surface of the atom, and renders it possible for the actual magnetism to assert itself and act upon another very near to it which it cannot have in any degree a little farther away, the actual distance being comparable with the diameter of the atom itself. Hence, atoms close by would have certain magnetic effects upon each other in the nature of selective effects, on account of the uniformity of the stress at the nodes, and the number of nodes would determine the possible number of cohesive attachments. So one may fairly presume that the vibratory motions such as constitute the heat motions of atoms are the physical conditions that underlie chemical combinations and give to them their quantitative character, their selective property, and their symmetrical form into which they arrange themselves.

This gives a rational account of so-called chemical attraction, and makes it clear how the laws of thermodynamics are related to chemical re-actions. It reduces the whole scheme to one of the mechanics of vibrating magnets; and the evidence that atoms are such magnets does not rest upon the necessity of the conception for the hypothesis, but upon much confirmatory experiment that has led physicists to the conclusion that they are such, in a manner quite independent of what phenomena might be deducible from matter with such a constitution. In conclusion, it may be added that, although the idea of ring-formed atoms has been adhered to in this explanation, it is not to be understood that the same explanation would not apply to atoms constituted in any other manner; for all that is implied in the above is that whatever their form and substance they are magnets, and that they are so elastic as to be capable of internal vibratory movements—that is, of changing their forms in a periodic way; and of this there appears to be no reasonable doubt. When several such are combined together the resultant motions and their effects become very complicated, and therefore difficult to disentangle; but that would be no reason for not holding a well-grounded conviction that all chemical phenomena are truly physical, and referable to fundamental mechanical laws, and are fully explained when these mechanical conditions are pointed out.

CHAPTER X

Sound

THE term "sound" has two very different significations,—one a physiological one referring to a sensation in the organ of hearing, the other the physical cause of the sensation. When one has the sensation of sound, of course he usually infers that it was caused by some external physical condition that has in some way impressed itself upon his auditory apparatus; and, to one who has thought but little about it, it is difficult to get rid of the idea that sound is a something which exists, whether it be heard or not. That is, there would still be sound though there were no ears, that a tumbling pile of books in a deserted house would make a racket if no one did hear it. On the other hand, one may call that sound which is capable of being heard; and when those conditions are investigated it is found, in all cases, to be some kind of a mechanical impulse, or succession of impulses, generally in the air, which may be traced from the ear to some body which is found to be in a state of vibration. The latter is called a sounding body, and the air is called a sound conductor; but these conditions are not necessary for the sensation of sound. One may not infrequently hear what is called ringing in the ears, that has its origin within the head, and, perhaps, in some cases independent of any of the auditory apparatus, like some nerve disturbance even at the base of the brain itself. Hence there is a distinction between hearing and the cause of hearing, and the latter does not necessarily imply anything external to the listener. One may be deaf so that no conditions external or internal will produce the sensation. As the sensation itself can give no infallible testimony as to what causes it, it has come about that the physical conditions which may be heard as sound have been investigated, and the science of sound, or acoustics, has been developed quite independent of the sense of hearing, the latter being only a convenient instrumentality in the investigation, not an indispensable one. In this sense sound is the science of the vibratory movements of elastic bodies, and one may inquire first as to the origin of such movements. When one body strikes upon another, motion is imparted to the latter. If enough motion is imparted, it may move visibly, and we then call such motion mechanical. Though it does not visibly move, yet energy has been spent upon it in some degree, and must be represented by some degree of motion which at first it did not have. If a pencil be struck upon the table, one may be as sure that energy has been spent upon it as if it had been struck with the fist, only less in amount.

When molecules are compressed together so as to increase the density, and retained in such closer compactness, heat is always the result; that is, the molecules themselves have their amplitude of vibration increased: but when molecules are compressed quickly, and the pressure be as quickly removed, the compressed molecules at once rebound to their original position with a velocity that depends upon the degree of elasticity the body has, and, like a swinging pendulum, do not stop at once when they have reached that position, but go beyond a little, and thus oscillate back and forth. Each molecule pushes against its neighbors, and they upon theirs, and so on, the motion travelling outwards from the point of disturbance in every direction, with a velocity that is proportionate to the temperature; that is, the vibratory rate of the molecules themselves, which, as pointed out in the chapter on heat, is exceedingly great.

This particular kind of movement is called longitudinal; that is, it is to and fro in the direction in which the disturbance travels, and depends altogether upon the properties of the body that is struck, and not in any degree upon the initiating cause. When the table is struck with the pencil the sound heard is different in quality from that given out by a similar stroke upon the window or a tumbler. It differs also in duration. The latter may continue to be heard for some seconds, while the former is brief. Every elastic body has some particular vibratory rate, which depends upon its size and shape as well as the material it is composed of. A stretched string or wire, a board, a lath, a bridge, a house, for examples, all have individual rates of motion, into which they can be brought by some well-directed, sudden push. When a strong wind shakes a house, the shake is the vibratory rate of the building, and may be as low as one or two per second. In general, as bodies are smaller their rate of vibration increases, until it becomes greater than thirty or forty per second, when the effect can be heard. Stones have an individual pitch, or rate of vibration, so that by selection one may get a set to represent the musical scale when struck. Smooth bits of laths of different lengths give out their pitch when dropped upon a table; and, with a properly graded set, tunes may be played by dropping them successively. The rate of vibration, or pitch, of a table is relatively high—several hundred per second; and a pencil knock distributed over so large a body, and by it to the floor, reduces its strength very fast.

The tumbler has its motions symmetrical, therefore of greater amplitude, and last longer. A tuning-fork struck and held in the fingers near the ear will be heard for a much longer time than if the stem be held against the table, as any one may satisfy himself by trying. In the latter case the motions are conducted away freely, in the former case not so freely. In the former case the sound appears louder to the ear, because the air, in contact with the vibrating table, receives vibratory motions from it as well as directly from the fork; and so the air motions are re-enforced, and the energy is dissipated so much the more rapidly.

The idea in all this is that, so far as sound consists in vibratory motions, energy is involved, and is distributed in accordance with mechanical laws; the size, density, and elasticity of the sounding body being the factors which determine the rate at which the distribution can go on.

If the motion be properly mechanical, any agency that can originate such motions can give rise to sound. One might ask himself here if it be likely that any kind of motion, or form of energy, cannot produce it. If it be remembered that motion is the antecedent of motion in all known cases, one will perceive that sound might have a variety of antecedents, as it has. To the mechanical ones alluded to might be added all cases of percussion, impact, friction—indeed, the whole range of mechanical motions. Any agency that can change the form of a body can cause sound vibrations.

That heat can directly produce sound is shown by the roar of fire in furnaces; and tubes having a burning gas-jet in them may give out a loud sound. In these cases it is the body of air that is caused to vibrate energetically.

When a beam of light falls upon a body that can be heated by it there is a re-action between the surface and the air, in which the surface is pushed slightly backwards, as indicated by the radiometer. If a beam is allowed to fall intermittently upon such a surface, it will be thrown into vibrations as if it had been struck, and will give out a sound, the pitch of which depends upon the number of interruptions per second. Such a device is called a radiophone.

A current of electricity sent through a conductor in an interrupted manner makes the wire give out a sound. The current heats the wire, expands it slightly, and cools as suddenly when the current is stopped; so the succession of currents results in sound. In like manner, a current of electricity going through an electro-magnet causes a click at the instant of making and breaking the current. This is occasioned by the change in position of all the molecules. A succession of these may keep up a continuous hum.

The electric spark itself always produces a snap of brief duration, for short sparks from induction coils and electric machines; but, when the spark is a long one, like a flash of lightning, the sound may be prolonged several seconds. Along the line of the flash the air is greatly heated for a very brief time, and it therefore rapidly expands. The quick cooling produces a collapse of the heated column of air, with the consequent noise. The duration of the thunder does not signify that the lightning lasts such an appreciable time, but that a part of it was a distance away, and that time was taken for the sound to come from the more distant place.

That chemical action can give rise to sound is proved by the explosion of gunpowder and other explosives, solid and liquid. In these cases a large amount of gas is suddenly formed, and at a high temperature; it displaces the air quickly and forms a great wave. One may often feel the wave of compression produced by a cannon go by him, even at the distance of several hundred feet from it. These examples show that heat, light, electricity, magnetism, and chemism are directly related to mechanical motions because competent to produce them under appropriate conditions. If motion be the antecedent of any given motion, and any of these may be the immediate antecedent of mechanical motions such as sound, what shall be said as to the nature of each of these physical agencies?

CHARACTERISTICS OF SOUND.

As sounds may be produced by any of the physical agencies, it does not matter, except for convenience, what ones are adopted. Usually mechanical motions are most convenient, and for musical purposes either percussion, or currents of air. We speak of high sounds and low sounds, and we find by experiment that those called low are produced by fewer vibrations per second than those called high. If sounds are considered as vibratory movements, then it is evident there is practically an infinite range of them; for there may be any rate, from one a year or a thousand years all the way to such vibrations as atoms make, measured by millions of millions per second. There is no good reason for drawing a boundary-line at one point rather than at another, and saying that all vibratory movements beyond this rate are not to be considered as sound, yet it is convenient for some purposes to confine the range to such as can be heard.

When a succession of impulses follow each other at such a rate as just to produce a continuous sensation of sound, it is found to require from twenty to thirty per second. It differs very much in individuals. In the young it requires more, as the organ of hearing acts more promptly than it does in the old. A less number than these is heard as a tremble. From this as a minimum one may go through a series, running from the lowest sound produced by a piano—about forty per second—to the highest one of about 4,000 per second. Many insects make much higher sounds than this. Such differences in the rate of vibration are called differences in pitch; and, for musical purposes, a standard of pitch has been adopted, making the middle C of the piano give from 256 to 261 vibrations. The pitch of a sound may be specified by giving its vibratory rate. The pitch of men's voices ranges from 100 to 150 vibrations in conversation. Ordinary whistling is produced by from 1,000 to 3,000 or 4,000. The squeak of bats is in the neighborhood of 5,000. Beyond these figures it is difficult to hear anything,—not because the vibratory motions are not produced, but because they have too little energy to affect the ear. Occasionally aurists find abnormally sensitive ears capable of hearing sounds with a pitch as high as fifty or sixty thousand, but ordinary persons have a limit in the neighborhood of 20,000; so it is customary to say that the range of

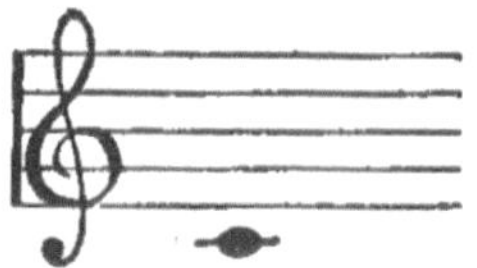

DIAG. 33.

hearing of mankind is from thirty per second to about 25,000: but it should always be borne in mind that the chief reason for not having a greater range is in the difficulty of giving sufficient amplitude to such very rapid changes. As the pitch rises the amplitude decreases for a given amount of vibratory energy. One might attribute the relatively low vibratory rate of the maximum which the ear can perceive to the lack of delicacy of the apparatus itself, which would be true enough in an absolute sense; but the actual sensitivity of the ear is really something wonderful, for a piece of apparatus that is altogether mechanical in its mode of operation. It has been found that the ear can hear such sounds as are produced by small whistles at the distance of several hundred feet; and, if the amplitude be computed,—assuming that it varies inversely as the square of the distance—it is found to be comparable with the diameter of a molecule, or less than the ten-millionth of an inch. One who understands the necessity for vibratory motions in elastic matter will readily conclude that between the highest number the ear can perceive, say 50,000 per second, and the lowest rate capable of affecting the eye (400 millions of millions), there is an enormous gap; and man has no organs for perceiving the intermediate ones.

Experiments made in various ways have shown

that the velocity of sound waves in air is about eleven hundred feet per second, and varies with the temperature, being only 1,090 feet at the freezing point of water, increasing or diminishing about two feet per second for each degree above or below that; and this is true for sounds of all degrees of pitch. If it were not so, music could not be heard at any distance from its source. Suppose a tuning-fork makes one hundred vibrations in a second. At the end of the second the first wave would have got say eleven hundred feet away, while the last wave would have just been completed; or between the fork and the more distant wave there would be a series, one hundred in all, reaching eleven hundred feet. It follows that each wave would be eleven feet long, or the velocity of transmission divided by the number of vibrations. The wave length of sounds can be measured in several ways, and of course the product of the wave length into the number of vibrations gives the velocity of sound in any conductor. An idea of the actual wave length for common sounds may be had thus: If the middle C of the piano makes 261 vibrations per second, and the velocity in the air of the room be 1,140 feet, $\frac{1140}{261} = 4.36$ feet, as the length of the air wave, and for a man's voice it will be about $\frac{1140}{125} = 9.1$ feet, while the highest note on a piano will be $\frac{1140}{4000} = .285$ foot, or 3.4 inches. In water the velocity is four times greater than in air, in wood about twelve times, and in steel about sixteen times greater; and this will give a corresponding increase in the wave length. This velocity of sound in air is, roughly, about a mile in five seconds, or twelve miles a minute; and at this rate nearly a day and a half would be needed to go round the earth.

Air waves, like water waves, are reflected when they come against a more solid body. Such reflections of air waves are called echoes. The mere fact of reflection does not change the length of the wave, as the pitch of a sound is not altered by having its direction changed. The law of sound reflection is the same as that for the reflection of energy in general; viz., the angle of reflection is equal to the angle of incidence. Neither does reflection change the velocity of sound waves.

The phenomena of echoes are familiar to every one, for walls, houses, wood, and hills all echo sounds; and one may roughly determine the distance to such an echoing surface. As one approaches such surface the time between producing a sound and its return is shortened, until, when about sixty feet from it, the two so blend that the echo is no longer heard with distinctness. The sound has then travelled 120 feet.

When sounds are produced at the ends of tubes the walls of the tube prevent, by reflection, the scattering of the waves, and the whole motion is kept in nearly parallel lines, and with slight loss in strength; hence the utility of speaking-tubes. If the tube be a short one, and stopped at one end, a new phenomenon appears for sounds having a wave length about four times the length of the tube. The sound is much strengthened. A tuning-fork making say 435 vibrations per second will have a wave length of about thirty-one inches. If it be held while it is vibrating over a tube or vessel of any sort, between seven and eight inches deep, the increase in the strength of the sound will be very marked. The motion in the air is so much swifter than the prongs of the fork that, while one prong is beating downwards and thus producing a condensation in the air, the wave reaches the bottom of the tube; there it is reflected, and gets to the top just as the prong of the fork has returned to its normal position. As the fork continues upward, forming a rarefaction, the rarefaction also travels down the tube, and is reflected so as to get back when the prong has returned to its normal position; so for a complete vibration of the fork the air wave has travelled four times the length of the tube. It is possible in this way to make quite accurate measurements of either the wave length of a sound, its velocity, or the number of vibrations a sounding body makes per second. This phenomenon is called resonance; and it is the chief factor in wind musical instruments, such as flutes, organ-pipes, and the like. Resonance in general means the ability of a body to be thrown into sound vibrations by sound waves, and there are two well-marked cases that need to be considered. When the stem of a vibrating tuning-fork is held upon a table the sound in the air is much louder, for the whole table is made to vibrate at the same rate as the fork. The table will resound loudly to forks of any pitch. Such vibrations as are different in pitch from that belonging to the body itself are called *forced* vibrations. Resonance of this sort is the function of the sounding-boards of pianos, the bodies of violins, guitars, and other similar instruments.

If two tuning-forks have the same pitch, and one of them be made to sound, the other one will presently be made to sound also, though it be several feet away from the former one. The air waves act upon it like a pusher upon one swinging; at each return a little more energy is added, until the amplitude has become great enough to make the sound audible. Such vibrations are called *sympathetic,* for they are only effective upon bodies whose own rate of vibration is the same as that of the sounding body. Raise the damper to the piano and sing a sound of any particular note, then listen. The same note will be heard prolonged by the piano. The particular string which can give that pitch of sound has been thrown into similar vibrations, and continues to sound as it would if caused to in any other way.

The air as a body is too large to have a vibratory rate of its own, and, consequently, all sounds in it are properly called forced vibrations; but, when it is confined in cavities, resonance becomes apparent, and sympathetic vibrations may be so strong as to be deafening. That is the case often in locomotive furnace-flues when the door is opened. One may hear it a mile or two. The resonance of large rooms sometimes renders it very difficult to understand a speaker in them.

The prolonged sound of thunder has been often explained as due in some measure to echo from the clouds, but it is doubtful whether clouds do echo sounds. No one ever hears the sounds of bells, whistles, or cannon, or other strong sounds, coming from the clouds, as would be the case if they reflected sounds appreciably.

When a single key of a piano is struck, there is produced what is called a musical sound. There is a definite pitch that is maintained. Strike half a dozen adjacent keys at once, and the effect is what is called a noise, though each component by itself would give a pleasing sound. A load of stones when tipped from a cart makes a great racket; yet each stone, if struck with a hammer, may give out a distinct musical sound. Nearly every body has its own musical pitch; but, if a number of bodies with different unrelated pitches are listened to at once, the effect upon the ear is a discordant one, and is called a noise.

When, however, two or more musical sounds whose pitches stand in a simple ratio to each other are heard together, they blend so as to form a pleasing combinational sound. Thus, if one makes twice as many vibrations per second as the other, the sound is a very smooth musical one, and one is said to be the octave of the other. If middle C of the piano makes 261 vibrations, the octave above will make 522, and the octave below 130.5; and these may all be heard at once as a musical sound. In music an octave is divided up into eight parts called tones; and these are sung as *do, re, mi,* and so on. If a string be stretched between two points and the distance measured, the sound it will produce may be called *do* of the scale. If the string be now shortened by a bridge so as to produce the note *re,* and the length of the string be again measured, its length will be found to be eight-ninths of the length of the first, the note *mi* will be four-fifths, *fa* three-fourths, *sol* two-thirds, *la* three-fifths, *si* eight-fifteenths, and the next *do* one-half. As the number of vibrations a stretched string will make is inversely as its length, it follows that these fractions inverted will represent the relative number of vibrations produced by each member of the musical scale when compared with the beginning or fundamental one. The following shows the letters of the musical scale, with their ratios and vibration numbers for the middle octave of the piano.

C	D	E	F	G	A	B	C
	$\frac{9}{8}$	$\frac{5}{4}$	$\frac{4}{3}$	$\frac{3}{2}$	$\frac{5}{3}$	$\frac{15}{8}$	$\frac{2}{1}$
261	293.62	326.25	348	391.5	435	489.37	522

The meaning of this is that $\frac{9}{8} \times 261 = 293.62$, and so on, so that the notes of the musical scale stand in simple ratios to each other; and, if one has the vibration rate of any one of them, he can compute any others. Of course any octave above this one will have simple multiples of these numbers for their vibration numbers.

But these numbers signify more than simply this: they signify that, when a second one is sounding with C, it will make the number of vibrations represented by the numerator of the fraction; while C is making the number indicated by the denominator. Thus, G makes three vibrations while C makes two. The sounds are concordant one-third of the time, and the effect is a pleasing tone. On the other hand, D makes nine while C makes eight, and the two are in accord but one-eighth of the time; and the effect is displeasing, and is called discordant. The smaller the ratio the more musical and harmonious the sounds; and music is made up of a succession of sounds standing in such relations to each other, and, when different ratios are employed, it is only for contrast, and return is quickly made to these ratios. The ear will not long tolerate a departure from them.

It has been stated that sympathetic vibrations would cause a given body to vibrate. Press down gently a base C on a piano, so as not to make it sound. Now strike the C above it, holding down the key for a second or two. On letting up the latter the sound of the latter will continue to be heard, but coming from the lower key, as can be learned by letting up the key, when it will cease to be heard. If the G above the struck C be now struck with the same low C held down, the sound of the G will be heard from the base string, and so one may go up, finding eight or ten strings, each one of which will make the low C string vibrate, giving out the sound of the higher string. It is found that each one of the strings able to do this has a vibration number which is a simple multiple of the lowest one. The first one is the octave, making twice the number; the second one is the fifth of that octave, making three times the number; and so on, to the upper limit of the piano.

This means that a piano string is capable of vibrating in a number of rates,—two, three, four, and so on, times its own lowest rate, which is always called its pitch. It is also found that this process is reversible; that is, if each one of these keys in turn be held down and the lowest one struck, they will each be set vibrating; and this shows that the struck string vibrates itself in the several different pitches represented by the multiples of its fundamental rate. The sound of a piano string is therefore a compound sound. In such a compound sound the lowest one is called the fundamental, and the others the over-tones, or harmonics. Some of these harmonic sounds are likely to be stronger than others; and some may even be so much more energetic than the fundamental as to nearly drown the latter, so as to make the pitch of the string to appear an octave or more higher than it really is. The number and relative strength of the harmonics in a compound sound make the difference in the quality of sounds. In all such instruments as pianos, violins, guitars, and the like string instruments, the number and strength of the over-tones depend in a large measure upon how and where the strings are struck and made to sound. A piano string plucked near its middle point gives a different sound from what it will give if plucked near one end, and different in each case if plucked by the fingernail and by the finger. So the quality of sound can be much modified by mechanically varying these factors.

In other musical instruments the sounds are also compound in a similar way, differing in the number and strength of the higher harmonics. Some have the even harmonics, as the second, fourth, sixth, and so on, stronger; some have the odd ones—first, third, fifth, etc.—stronger; some have few, and some many. A flute has but one or two, a violin has twenty; and thus the character of the sounds of musical instruments is explained.

As for the voice, the sound is produced by the vibrations of what are called the vocal chords, which are fixed at the junction of the trachea and æsophagus, and through which all the air to and from the lungs has to go. These chords are modified in tension by muscles at will, and so change the pitch of the vibrations. The cavities of the throat, the mouth, and nose act as resonators for these sounds and seem to strengthen some of the constituents, thus giving prominence to certain ones to the exclusion of others. That the mouth acts this way may be observed by pursing the lips as if to produce the various sounds of ah, oo, o, snapping one cheek with the finger. These sounds will result; while, with a little trial, one may thus snap a tune which may be heard through a room, merely altering the size of the mouth cavity. The cavity of the nose is as important as that of the mouth. When this cavity is small and narrow, there is produced what is called a nasal sound. When this is prominent, and is not the result of a cold, as is sometimes the case, the trouble is a physiological one, due to the bad shape of the resonating cavities rather than to careless habits, as is often assumed by some teachers of expression. Some different pitch of the voice in ordinary speaking might be adopted, and thus in some measure prevent the disagreeableness of the nasal sounds, but no amount of painstaking can altogether prevent it. That structure and its acoustic effects are an inheritance in some parts of the world, as are crooked noses, thick lips, black eyes, and broad heads in other and different parts of the world, and is no more to be legislated away than are these other physiological peculiarities. Neither is it a proper subject of ridicule, more than lameness or defective vision.

If the bell of a locomotive be rung while it is swiftly approaching one, the pitch of the sound rises until the engine has reached the observer. As it retreats the pitch lowers, and the difference in pitch becomes greater as the velocity of the engine is greater. The explanation of the phenomenon is, that one judges of the pitch of a sound by the number of vibrations that reach the ear per second. Suppose an observer be distant eleven hundred feet from a source of sound of one hundred vibrations per second. If both observer and source remained in place, one hundred vibrations per second would reach the ear of the observer, and there would be one hundred more on the way to his ear. If the observer should continue to go that whole distance of eleven hundred feet to the source of the sound in one second, he would not only receive all he would by standing still, but in addition all that were on the way to him,—two hundred vibrations in all,—or just twice the number that would reach him if he remained in place. Now, twice a given number of vibrations represents a difference in pitch of an octave. The sound he would hear would be an octave higher than the sounding body was actually making. Any less velocity than that supposed would make a corresponding less difference in pitch, but such velocities as railway trains have may make a difference in the pitch of more than a musical tone. Of course, if the sounding body and listener be separating, a less number of vibrations will reach his ear, and the pitch will be correspondingly lowered. One may roughly determine the velocity of a train of cars by noting the change in pitch of bell or whistle. Thus, if the difference be, say a musical semitone,—one-sixteenth,—then the speed of the train is one-sixteenth the velocity of sound in air, one-sixteenth of 1,125 feet, which gives seventy feet per second, or forty-seven miles an hour.

The ear is a complicated structure of tubes, muscles, cartilages, bones, fibres, and nerves. The external part, or conch, is of but little service in hearing in man, for it cannot be directed, as can the ears of horses and cattle. If it stands out from the head so as to have some use as a collector, it is supposed to be in abnormal position; but it is not much needed in any case. The orifice of the ear is known as the tympanum, a tube a little over an inch in depth and about a quarter of an inch in diameter. At the inner end it is covered with a thin membrane called the drum of the ear. On the inner side of this membrane, there is attached to the middle of it a bone fixed to a kind of hinge, so that any movement of the drum of the ear, in or out, makes this bone to move in a similar way.

Then follows a network of bones and cartilages, and a set of fibres known as Corti's, of different lengths, and whose function has been supposed to be for sympathetic vibrations. There are in the neighborhood

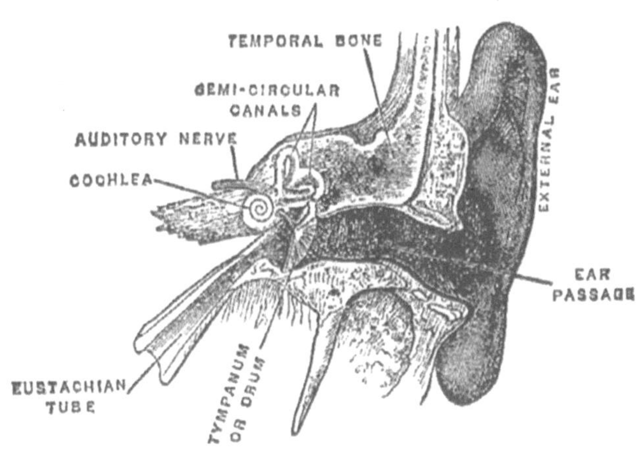

DIAG. 34.—THE EAR.

of 4,000 of these fibres, each one adapted to vibrate at a different pitch. Then follow the nerve terminals and the acoustic nerve itself, which goes to the base of the brain, where its function as an acoustic instrument ends with the delivery of its peculiar motions, interpreted by consciousness as sound.

It is easily seen that the whole structure is one adapted to receive vibratory motions from the air, within prescribed limits, and transmit them inwards where they can be interpreted. The tube itself possesses resonating properties like any other tube. The membrane is shaken to and fro by sound vibrations, and this movement is handed on to each distinct part until the nerve itself is shaken. From beginning to end, it is only the transfer of a particular kind of motion,—what is called mechanical,—perhaps transforming it from longitudinal to transverse vibrations. That it is so extremely sensitive as to be affected appreciably by motions so slight as the ten-millionth of an inch is a marvel, and shows that mechanical motions of translation, though on a scale of molecular magnitudes, is able, through the proper avenue, to affect the mind and develop consciousness, which experience enables the individual to interpret by direct inference.

Let one reflect upon the facts furnished in great abundance by physical science,—that all the data which comes to the mind through consciousness, and which furnishes what is called experience, is simply motion of some sort. Touch, producing pressure upon the surface of the body, finds a suitable nerve to transmit to the base of the brain that kind of a disturbance; sight, another kind of disturbance to the optic nerve, transmitted to the same place; hearing, still another kind of motion given to another kind of nerve running to the same headquarters. So, by means of motions of various sorts man determines his place in the universe, and learns how he may adjust himself to it.

CHAPTER XI

Life

ANY scheme of physics which fails to present that great body of physical phenomena exhibited by living things, both vegetable and animal, must be incomplete. Many of these phenomena have seemed to be so remote from ordinary mechanical operations that, in the absence of definite knowledge concerning them, their origin, factors, and relations to subsequent phenomena, it is not to be wondered at that they were long thought to be due to some peculiar force residing in a living thing, which was not to be attributed to the general endowments of matter, but only to be found in certain organized forms of matter, which organization it had itself built up as a *habitat*. It was conceived to exist apart from any material organization as a kind of entity. The difference between a living and a dead animal was thought to be simply one of the presence or absence of that entity called life. It was thought to be able to effect changes in matter which the ordinary physical and chemical forces could not possibly do; and many of the chemical products of living things were supposed to be formed only through its agency; and still more than that: it was held to be capable of "suspending the action of chemical laws." That the stomach itself was not digested by the gastric juice it secreted was held to be proof of its control over chemical operations.

There have been many attempts to define life, but the efforts have not been very successful. Thus Kant defines it as "an internal principle of action;" Treviranus, "the constant uniformity of phenomena under diversity of external influences." Bichat, "Life is the sum of the functions by which death is resisted." Duges calls life "the special activity of organized beings." De Blainville's and Compte's definition runs thus: "Life is the twofold internal movement of composition and decomposition, at once general and continuous;" and Spencer's is "the continuous adjustment of internal relations to external relations." It will be observed that in all of these what is described is a series of processes, or a body of functions belonging to certain structures, rather than an entity,—a description of what life does rather than what it is.

Analogous difficulties were met in the attempts to define other of the so-called physical forces. Thus light was supposed to be a created something. The corpuscular theory of it represented it as consisting of particles of some sort that ordinary matter could absorb and eject, and which, therefore, had an existence independent of matter. The establishment of its being but wave motion in the ether completely destroyed the notion of its having an objective, independent existence.

Heat, too, was supposed to be a kind of imponderable matter, and certain phenomena in ordinary matter depended upon its presence or absence; it, therefore, was supposed to be an entity, and to have an independent existence. Experiment showed it to be but a particular kind of motion, so the idea that there was any such thing as heat was abandoned.

Electricity and magnetism were supposed to be fluids; and some of the early terminology still survives in popular speech to-day, as when one reads that the electric fluid struck a tree or entered a house. Nevertheless, nobody now believes that either of them is a fluid, or has an existence independent of matter.

The regular movements of the planets were thought to require intelligent directive power to keep them in their orbits; but now the gravitative property of matter itself is held to be quite sufficient to account for all the observed facts, and the extra material directive force is held to be an entirely unnecessary assumption.

The discovery of the conservation of energy, covering every field that has been investigated, led to the growing conviction that there are no special forces of any kind needed to explain any phenomena. What seemed probable forty years ago, to those who were conversant with the facts,—that vital force as an entity has no existence, and that all physiological phenomena whatever can be accounted for without going beyond the bounds of physical and chemical science,—has to-day become the general conclusion of all students of vital phenomena; and vital force as an entity has no advocates in the present generation of biologists.[1] The term has completely disappeared from the science, and is only to be found in historical works; and every phenomenon which was once supposed to be due to it is now shown to be due to the physical properties of a particularly complex chemical substance known as protoplasm, which is the substance out of which all living things, animals and plants, are formed. This protoplasm is entirely structureless, homogeneous, and as undifferentiated as to parts as is a solution of starch, or the albumen of an egg. Minute portions of this elementary life-stuff possess all the distinctive fundamental properties that are to be seen in the largest and most complicated living structures. It has the power of *assimilation*,—that is, of organizing dead

[1]See Appendix, p. 477.

food into matter like itself,—and, consequently, what is called growth. It possesses the ability to move—that is, of visible, mechanical motion, which is technically called *contractility;* and it possesses *sensitivity*—that is, ability to respond to external conditions.

It was formerly thought that the cell was the physiological unit, a cell having walls differently constituted from the substance enclosed, also a nucleus; but as the microscope was improved, and anatomical research continued, it became evident that the cell, with its more or less complicated structure, was itself built up by the structureless protoplasm. As before stated, it is a highly complex substance, chemically considered, made up of many atoms of carbon, hydrogen, oxygen, and nitrogen, with a small number of atoms of sulphur and phosphorus,—more than a thousand of them in one molecule; and there appears to be a great number of varieties of it. A small pellicle of this substance, like a minute bit of jelly, without any parts or organs, possesses its various attributes in equal degree in every part. Any particular portion can lay hold upon assimilable material, or digest it, or be used as a means of locomotion; so that what are called tissues of animals and plants are only the fundamental properties of the protoplasm out of which they have been built—thrown into prominence by a kind of division of labor. The protoplasm organizes itself into cells and tissues in the same sense as atoms organize themselves into molecules, and molecules into crystals of various sorts, having different properties, that depend upon the kind of atoms, their number and arrangement in the molecule.

The greater the number of atoms in a molecule the less stability does it have, and especially is this the case with molecules containing nitrogen. Many of its compounds are so unstable as to be liable to explosive disruption. This fact makes it easy to understand how there exists, in a mass of such molecules no larger than the minute ones seen in the microscope, conditions for internal motions in the nature of explosions.

Let it be granted that atoms are in the neighborhood of the fifty-millionth of an inch in diameter; then, if a thousand of them are organized into a molecule, its diameter would be about the five-millionth of an inch. A speck of protoplasm, one ten-thousandth of an inch in diameter, would require not less than five hundred such molecules in a row to span it; and there would be no less than one hundred and twenty-five millions of such molecules in the small mass. Some of these molecules would be less stable than others on account of the internal motions that all the time are present. Physical disturbances, external to such a mass, such as temperature, ether waves of light, and chemical re-actions of any sort, and so on, can induce and add to the disruption and other changes going on, and visible motions might be expected to follow.

That such external agencies can bring about visible motions of microscopic particles has long been known. A few small bits of camphor dropped upon the surface of clean water in a saucer will begin to move about in a remarkable way. They will spin round, and travel from place to place, and dodge each other in a manner strongly like living things. A little gamboge, which is a reddish-yellow gum used as a pigment, if rubbed up in water and looked at through a microscope, will be seen to have its particles in constant motion like animalcules. This is known as the Brownian movement, and is caused by temperature changes between the particles and the water. Such phenomena are rather extreme cases of the re-action of external molecular conditions upon a small mass of matter, resulting in mechanical motions. In protoplasm there is added to these same external ones others of the nature of molecular explosions within the mass, and together they give rise to a number of effects, in which the transformed energy shows itself in redistributing the molecules, absorbing additional material, and movements of other sorts.

Biological researches within the past few years have added vastly to our knowledge of protoplasm and its properties; and there is no longer any question that its qualities are the expression of the various movements, chemical and physical, and belong to it simply as a chemical substance. Chemists have synthetically formed out of the various elements a vast number of substances that were not long ago believed to be formed only by living things; and there is but little reason to doubt that, when they shall be able to form the substance protoplasm, it will possess all the properties it is now known to have, including what is called its life; and one ought not to be surprised at its announcement any day.

Some of the phenomena exhibited by bodies called inorganic, such as minerals of many kinds, possess properties that are very like those supposed to belong solely to living things. A spider or a lobster will have a new leg or claw grow to replace one lost in any way. In like manner a crystal will replace a corner or side or any defacement so as to complete its symmetry before it will begin to grow elsewhere, and this in cases where the crystal has been defaced or incomplete for millions of years, as is found to be the case sometimes in geological specimens. Such phenomena have led some of the most thoughtful and best informed naturalists to query whether the evidence we have does not lend much support to the theory that *matter itself is alive,* and

The above picture is copied from a photograph. It represents the plume-like forms assumed by water when crystallized in a basin. The similarity it presents to vegetable forms is very striking. One may often see on frosty window-panes fantastic imitations of organic things which forcibly suggest vitality. They are too common to be considered coincidences.

that the difference we observe in things is simply one of degree rather than of kind. See opposite page.

In the brief space of this chapter, only an outline of the relations between vital and physical phenomena can be given, and of these, only a few of the more prominent ones. It will suffice to show that such phenomena as assimilation and growth, movement and irritability, or sensitivity, have antecedents of physical energy in the same sense as the movements of an electric motor have physical antecedents in electric currents, dynamos, steam-engine, and furnace.

The food of an animal consists almost altogether in highly complex molecular compounds. It may be said to be matter stored with energy. A pound of bread may have the mechanical equivalent of twelve thousand heat units, and if burnt in an engine would be better for heating purposes than a pound of coal. When this has been digested, and has done its work in the body, the excreted products are of course equal in weight to the original pound, for no kind of a physical or chemical process affects the quantity of matter in any degree; but the products themselves represent much less complex compounds, and the energy has been distributed through the body, carrying on its various operations. There is, first, that of ordinary movement, which can be measured in foot-pounds, as work of any kind may be. The blood in the arteries and veins has to depend upon a kind of hydraulic apparatus to keep it in motion. The temperature of the body demands a supply of heat measurable in heat units to maintain it, while the repair and waste going on through the whole body of all animals implies a distribution of the material necessary for the maintenance of the integrity of the tissues, as well as a separation and removal of the used-up material; that is, the material that has lost all its available energy. The energy for doing all this of course comes from the food, so the question is not as to its source and quantity, but it is, How is this transformation of energy in the body effected? Is it direct, or is it indirect? This is the same as asking as to the mechanism in the body, by means of which energy supplied is transformed to meet the various wants of the body.

Roughly, there are five different kinds of motion to trace the antecedents of in the body of any of the higher animals. First, there is the common mechanical motions of the bony framework, which transport the body from one place to another, or change the position of a part with respect to the rest, as when one moves an arm. Second, there is the motion of a muscle, wholly different in character from the first, for the shape of the muscle changes by contracting in length and increasing in diameter. The muscles are so attached to the bones that the contractions of the one cause the others to change their positions. The muscular contractions of the heart, arteries, and veins keep the blood circulating; and the same is true for the processes of digestion, breathing, etc.

Third, there is the motion constituting the temperature of the body, which, as has been explained, is altogether atomic and molecular in its nature, and is, therefore, in strong contrast with the other two.

Fourth, there is a kind of motion that is going on throughout the body of the nature of transpiration, in which solids, liquids, and gases are passing through the various membranes without rupturing them. In the lungs there is an exchange of gases, oxygen going one way and carbonic acid gas going the other. In all the mucous-membrane-lined cavities there is more or less liquid oozing through the walls continuously, and there is no tissue so dense but protoplasmic masses do not move into or out of apparently with ease. They go through the walls of veins and arteries as if the latter were porous bodies, though no visible pores have ever been discovered in them.

Fifth, there is the motion in the nerves, in the nature of a longitudinal wave, and the velocity of which is in the neighborhood of one hundred feet in a second, which, though it is slow compared with sound waves or light waves, is fast when compared with the other motions of the body. He is a swift runner who can run at the rate of thirty feet a second for any distance.

The contraction of a muscle is to be measured in fractions of an inch per second. The motion of heat, measured as a rate of conduction, is exceedingly small in the animal body,—probably not the hundredth of an inch per second. The transpiration, or osmotic action, is also a relatively slow movement, so that a velocity of one hundred feet per second, which is upwards of a mile a minute, is really rapid.

How and why the bone moves we know: it is because the muscle that is attached to it contracts; but how is energy spent to make a muscle contract? As a matter of fact, when a muscle contracts it evolves a considerable quantity of carbonic acid gas and water; it also becomes acid, all of which imply chemical actions, for these are chemical products. Carbonic acid gas and water are the chief products of the combustion of such material as foods, for they are made of what are called hydro-carbons (combinations of carbon, hydrogen, and oxygen chiefly); and when these elements re-combine, forming water and carbonic acid, there is always a relatively large but definite amount of energy given out in the form of heat, and this effect is independent of time or place; that is, the same amount is developed whether the process goes on fast or slow, or whether it takes place in a furnace, in the body, or by slow decomposition called rotting. When it goes on faster than the heat can be conducted or radiated away, the temperature rises and we say the body is hot. When the heat generated is at once employed to do work, as in a steam-engine, the temperature of it is reduced proportional to the work done. When this takes place in a contracting muscle better results follow, for conduction and radiation within a muscle can take place at only a slow rate; so the temperature rises, and this explains the sensation of warmth resulting from muscular exercise. The increase in perspiration is also partly due to the same re-action of decomposition, as water is one of the products. When the muscle in contracting does additional work, as in raising a weight, a corresponding amount of decomposition takes place, and the heat is but transient, as it is at once transformed into the muscular motion, which is as much mechanical in its nature as is the movement in a steam-engine.

The muscle is quite like a spiral spring, which may contract upon itself and do work by contracting.

It is not the substance of the muscle itself that undergoes the change of disintegration, evolving water, carbonic acid, and other products; but there is evidence that the muscle secretes a particular substance called *inogen,* the rapid decomposition of which causes the contraction. As this substance can only be replaced at a definite rate and in a definite amount, it is clear that the work of a given muscle is limited by the physiological processes that precede it. The rate of work of a muscle is then determined by the rate at which inogen can be secreted by the muscle, and work done beyond that rate results in muscular exhaustion, which in its early stages is called weariness, and requires repose for fresh accumulation. Excessive draught upon the muscles reduces their ability to secret inogen, and their degeneration follows.

Muscular contraction is satisfactorily accounted for without assuming any vital force. It has a purely physical origin, the structure itself acting as a kind of mechanism for transforming the chemical energy supplied in food into the mechanical forms of energy represented by the various movements of the body, external and internal, which have already been mentioned.

That physical and chemical agencies bring about new movements is of course well understood. Especially clear is this for such nerve actions as accompany the special sensations of sound, sight, touch, and the rest. That the disturbance is properly described as a movement is apparent when it is found that it has a rate of progression, as before stated, of from one hundred to three feet per second. Whether such movement be similar to a sound wave in a rod or tube, or to an electric disturbance, makes no difference so far as the transformation and transference of energy are concerned. For sound there is the antecedent of vibratory motions in the air; for light, waves in the ether; for touch, mechanical pressure; for taste, chemical solution; and for smell, gaseous substances with definite constitutions and rates of vibration. These represent the ordinary stimulants to action of such nerves, and so are commonly understood to be the source of disturbance; but every one of these so-called special nerves may be excited to action by other agencies than the common or normal ones, and the effect is the same. Thus, the optic nerve may be stimulated by pressure, by cutting, pricking, thumping, and electricity, and the effect is the sensation of light; and, in the absence of other sources of information as to the origin of the sensation, no one could tell which of these was the originating one. Every one of them, however, represents some form of energy spent upon the nerve. What is important to note about it is this,—the nerve transmits an impulse it receives, quite indifferent as to its source, and is interpreted as a definite sensation, quite independent of its origin. The latter is only an inference, and is, therefore, liable to be erroneous.

But there are several other kinds of nerves, each with some different function from the rest. Thus there are nerves running to muscles, causing them to contract, called motor, or efferent, nerves; secretory nerves, to glands that cause secretions; vascular nerves, that cause contraction or dilatation of the walls of blood-vessels; inhibitory nerves, that affect other nerves so as to moderate or entirely stop their action; reflex, or afferent, nerves, which convey disturbances to the brain or other nerve centres, but which cause no sensation; and still others known to exist, but the special functions of which are unknown.

To describe the action of any nerve is to describe the transmission of energy in greater or less amount, and transmission in all cases requires time. This does not mean that the energy which does the special work of moving muscles or the chemical transformations of foods into tissues is transmitted by the nerves, but that the transformations of energy already present in each place where the work is to be done are controlled by nervous energy in the same way as a local galvanic circuit is controlled by a relay, or the explosion in a mine is determined by an electric spark. The energy available for all the purposes of an animal, including man, exists in the material of the body. The activity of protoplasm in the various cells transforms the various food stuffs into the proper substances needed. The energy is already present; it is only differently distributed by protoplasm; and nervous action determines what changes, if any, shall go on at a given place.

Temperature determines whether any of the physiological process shall go on or not. Plants and animals of a low order, such as snakes, frogs, and fishes, may be frozen without injury. Some of the minuter forms of life can withstand arctic winters, for there is an abundance of insect life in those regions. On the other hand, a temperature of 140° is destructive to the life of everything except the seeds and spores of a few microscopic beings. Some of these have been known to survive a temperature of 200°, continued for an hour or more; but nothing has been found that can withstand the boiling temperature of 212°. The retarding influence of cold upon vital processes can be understood by considering that special chemical compounds require special temperatures to form; and, if energy has to be supplied to maintain the proper temperature, so much the less will be at disposal for other processes. If life processes were other than physical, it might be expected that they would not be quite so rigidly conditioned by physical surroundings.

There is a distinction between a living plant or animal and the seed or spore or egg out of which they grow. Both are commonly spoken of as living things, but the processes that constitute life in the one are not present in the other in any degree; thus, for example, growing corn and the grain of corn from which the plant started. The grain of corn may be kept in a suitable dry place for several years without any apparent change, unless it be some loss in weight due to evaporation from it. How long it may exist thus and still be able to grow if planted is not known. Grains of wheat found with Egyptian mummies buried three thousand years ago have been said to grow, but there is much doubt about it, and botanists do not credit the story. A few years' keeping in moister climates destroys their ability to grow, and farmers always choose seed corn from last year's growth, which is an indication that there is a process of slow deterioration going on that ends after no long time in utter inability to grow under any conditions. This ability to remain for several years in a nearly stable condition is a property of the seed that does not belong to the plant; for, when growth has once really begun, it must keep on growing or die: arrest is impossible, which seems to show that life is a process rather than a condition, and the grain of corn is simply a combination of materials where, under suitable conditions, life may begin.

The constitution of corn is well known; that is, the elements out of which it is built up, and the proportionate parts of each. Like other kinds of food, it has carbon, hydrogen, oxygen, nitrogen, for the chief constituents, and in addition a little sulphur, phosphorus, iron, potassium, and a trace of some others. These, when organized as they are in a grain of corn, form a very complex body indeed. There are not only molecular groups of many sorts, but these are segregated into families, so that bodies of one constitution are all in one locality, and bodies of other constitutions in other separate localities, but definitely arranged so as to be available when the life process begins. Once formed, it appears to be as inert as a crystal of any sort, and no change happens to it until such physical conditions as heat and moisture are provided. These it absorbs and transforms; a sprout appears, then a root, each with different functions, one for absorbing ether waves, the other for absorbing water. The energy of ether waves is utilized in digesting carbonic acid and building up the structure, and the growth is simply the addition of materials gathered in this way and elaborated into similar protoplasmic form and structure. Growth implies transformation of one substance into the material of another, and is effected by means of energy from external sources. The energy of a stalk of corn may be found by using it as fuel and finding its heat units per pound. It has about the same value as wood. The corn itself has somewhat higher value, which shows it to have a more complex molecular structure, and is correspondingly less stable.

In like manner an egg, say that of a hen, possesses a degree of stability that does not belong to it after it has begun to grow. It may be kept with some care for a few months and retain its ability to develop into a chick; yet it ultimately wholly loses its possibility, which shows that slow changes of the nature of disintegration are going on that cannot be arrested. The physical condition necessary to initiate the growth of the egg is simply one of temperature. One hundred and four degrees continued for three weeks completes the process. When one reflects upon the nature of heat,—that it is but vibratory motion,—he can at once see that energy has been supplied to a complex mass of matter and it has been chemically transformed. There are new chemical products and new properties produced; and however wonderful the completed product may be, the factors at work to produce it have been absolutely physical from beginning to end. After growth has once begun the process must continue, at the peril of quick degeneration on stopping; so that an egg, like the grain of corn, seems to be a material structure where life may begin, rather than a living thing itself. Such a distinction has not, however, been made in the literature of the development of living things. It has, perhaps, only a philosophical importance; but, if there are any who would still hold that life is a something *sui generis,* that may be considered apart from some material structure and not as a transformation process, it will be well for such to inquire what can become of such life as a grain of corn or an egg has when either of them is cooked, or when either of them is left for months or years and they rot. At first it is in the grain of corn or egg. If it be an entity of any sort it must be somewhere else after leaving either the one or the other. On the other supposition the question does not arise at all, for it is plain that disintegration destroys the molecular arrangement, and with the destruction of that the properties of such organizations of matter must go also; for the properties of a mass of matter are, by general agreement, the result of the arrangements of the matter. Woody fibre and starch are of precisely the same chemical composition, but the properties of the two are far from being identical.

What, then, is the distinction between what is called living and dead matter? One is able to transform energy for its maintenance, and the other seems to be wholly inert; yet, if analyzed, both may be reduced to precisely the same amount of elements.

An analogy may make the distinction plainer. A maker of physical instruments may make what is called a Toepler-Holtz electrical machine. It is composed of wood and glass and brass and tinsel and tin foil, and possibly of other materials. Each one of these is got at a different place from the rest, and all are assembled in the shop of the maker. The individual parts are shaped in particular ways, and these are at last fixed in their appropriate places. The machine is done; but it has never generated an electric spark, and one could discover no electricity about it. Indeed, there is none, any more than when the material was unshapen and lying upon his bench. If the proper kind of energy is spent upon it, however, it at once becomes electrified, and electrical energy may now be got from it in indefinite quantity, dependent wholly upon the proper turning of the crank. If that be turned the wrong way, or if it be stopped, the electricity soon quite disappears. Now, it is the function of such a machine to transform mechanical energy into electrical, and it does this so long as energy is furnished for transformation and the integrity of the machine is maintained. If one weighs the machine before it has been worked, and also while it is electrified, he will find no difference. If the brass buttons get off or displaced, if the belt gets broken or the glass cracked, the machine will weigh just as much as it did when they were in place; but the property of the machine to transform energy will be destroyed, and it may be as useless for the purpose as a coffee-mill would be. One might speak of the whole machine as an organism,—its wood and glass and brass as its molecular composition, its function depending upon each of these being in its appropriate place, and nothing more. It can only exercise that function when energy of the proper sort is turned into it. If its molecular composition is deranged in any of a dozen different ways, no one is surprised that it no longer responds to the turning of the crank. If the complete and perfect machine be called living, then the one with its parts disarranged so it can no longer perform its functions might be called a dead machine.

The egg may be likened to the machine. So long as its molecular arrangement is intact, so long it is competent to transform the heat supplied to it and exhibit new properties. When the molecular arrangement is interfered with, whether from within or without, its function as transformer ceases, and we call it dead.

It may be said, and often has been, that every

living thing has an ancestry of living things; and in human experience it is true. It is sometimes said that one cannot get out of a mass of matter what is not in it, which, in this case, might imply that matter itself is alive, as suggested a few pages back, though I have never heard any one so conclude. If one would apply this dictum, let him settle with himself before turning a new electrical machine whether the electricity he is to get from it is or is not in the machine, and how, if it be in the machine, he can get an infinite amount from it by simply turning the crank. He may reach the conclusion that what can be got out of a mass of matter depends upon its composition and structure.

In conclusion, one perhaps can do no better than to quote the words of Dr. Michael Foster, Professor of Physiology, University of Cambridge, England, as to the properties of protoplasm. "The more these molecular problems of physiology are studied, the stronger becomes the conviction that the consideration of what we call structure and composition must, in harmony with the modern teachings of physics, be approached under the dominant conception of modes of motion. The physicists have been led to consider the qualities of things as expressions of internal movements; even more imperative does it seem to us that the biologist should regard the qualities of protoplasm (including structure and composition) as in like manner the expressions of internal movements. He may speak of protoplasm as a complex substance, but he must strive to realize that what he means by that is a complex whirl, an intricate dance, of which what he calls chemical composition, histological structure, and gross configuration are, so to speak, the figures; to him the renewal of protoplasm is but the continuance of the dance, its functions and actions the transferences of the figures.... It seems to us necessary, for a satisfactory study of the problems, to keep clearly before the mind the conception that the phenomena in question are the result, not of properties of kinds of matter, but of kinds of motion."

If such be the case, it is clear that the solution of every ultimate question in biology is to be found only in physics, for it is the province of physics to discover the antecedents as well as the consequents of all modes of motion.

CHAPTER XII

Physical Fields

I.—THE THERMAL FIELD

WHEN a mass of matter of any kind possesses energy of such a kind as to be able to impart some or all of it to the medium about it, whether that medium be the air or the ether, which transmits or distributes it outwards with a velocity which depends solely upon the ability of the medium to transmit energy, and not upon the source of it, the energy so distributed is called radiant energy.

The term was first applied to the energy radiated by a hot or luminous body, from which the heat was said to be radiated away, the motions of the molecules of the hot body being transformed into wave motions in the ether. The wave motion thus set up is known to be competent to set other masses of matter upon which it falls into vibratory molecular motions, similar to those that originated the waves. In other words, they are capable of heating other matter. The space within which such effects can be produced will evidently be limited only by the distance to which the wave motion is transmitted, and this in turn depends upon the special medium concerned—in this case the ether—and the uniformity of its distribution. As has been already pointed out, the ether transmits such wave motions in straight lines, and to an indefinite distance,—so great at least as to require not less than five thousand years to cross the space accessible to our observations. As such waves of all wave lengths travel with equal velocities, and as all known bodies of matter are continually radiating waves of many wave lengths, it follows that in reality every molecule of matter sets the whole visible and invisible physical universe in a tremor. The magnitude of this effect is not now under consideration.

The space external to a body within which the body can act in this physical way upon other bodies, so as to bring them into a condition similar to its own, is called its *field*. The heat or thermal field of a mass of matter of any size and of any temperature must, therefore, be as extensive as the universe, unless the ether absorbs the energy to some extent and becomes itself heated. At present there is no evidence that such an effect is produced. Some astronomers have inferred that absorption takes place, else the whole surface of the sky would be bright with the multitude of stars that occupy it. On the other hand, if absorption did take place in a manner at all comparable with gaseous absorption, it would be selective in some degree, and the more distant stars would have a color different from those closer to us; and the colors of all stars would depend upon their distance from us. If such a condition had been observed, it would be conclusive evidence of absorption in the ether, but it has not been observed.

Furthermore, the perception of light implies a definite though a small amount of energy; and, as the energy of ether waves from a given point upon a surface varies inversely as the square of the distance from the point, it follows that there must be some distance from it where the energy upon the retina must be too slight to affect it; and hence the inability of the eye to perceive the light could not be used as an argument against the existence of the waves altogether. At the rate of 186,000 miles per second light travels 5,800000,000000 (nearly six millions of millions of miles) a year, and in five thousand years, which is the distance of some of the more remote stars, 29000,000000,000000 (twenty-nine thousand millions of millions) of miles. This, therefore, is the known length of the radius of the thermal or light field of a heated or luminous body; and, as such heat-producing waves are radiated in every direction about the body, the sphere having such a radius represents the space within which any or every atom of matter can affect other atoms to heat them.

II.—THE ELECTRICAL FIELD.

The phenomenon called electrical induction, by which one body becomes electrified by simply being in proximity to another body which is electrified, is another illustration of both a *field* and its property, depending altogether upon its origin. But an electric field differs in a marked way from a thermal field.

Imagine a sphere—say a cannon-ball—to be electrified, and be isolated a long way from any other body. Its effect upon the ether about it would be equal in every direction. Practically, it would be distributed as the thermal field would be; and, if the strength of the field should be measured in any way, it would be found to vary inversely as the square of the distance from the body that produced the field. When such an electrified body is adjacent to other bodies, as is necessarily the case with every electrified body upon the earth, the strength of the field at a given point is found to depend upon the size, the nearness, and the quality of the adjacent body. Suppose the adjacent body were a similar cannon-ball, and its distance from the former one foot. Then the strength of the field would be found to be greatest between them, and to be very weak in the space equidistant and on the opposite side. One may get a mechanical idea of the condition of things by imagining straight lines drawn from the electrified body when out in space as if they were rays of light, evenly distributed in space. When, as in the second case, another ball is near to it, these rays crowd around the second one and apparently are absorbed by it; and these may now be represented by the same lines, starting at the same places as before, but sweeping in curves to the second, with only here and there one to escape into the unoccupied space. The nearer the two are together the more closely are these lines crowded together in the space between; and, as the number of these lines in a given area represent the strength of the electric field, it is plain the field is strongest where the lines are most crowded. On the other hand, if the second ball had been made of glass, the field would have been changed but little, for glass is a substance having but little absorptive power for electric rays; that is, it is not much affected by an electric field. When such an electrified ball is suspended in an ordinary large room, these lines, representing the field, are distributed about the room in a manner that depends altogether upon the kind of material there is in the room. The metallic objects, such as a stove, a steam-radiator, a gas-pipe, and the like, will divide the field between them, not equally, for the nearer ones will have the most, and other parts of the space in the room will have but a trace of it. The great distinction between the electrical and the thermal field will be apparent when one reflects upon what the latter would be for the same cannon-ball made hot and suspended, in the same manner, in the room. The rays go straight in every direction, and are not deflected by proximity to other bodies. The one is uniform in every direction about it; the other is warped by the presence of other bodies.

An electric field, which is merely the ether in a condition of stress, electrifies the bodies upon which it acts; that is to say, it produces in them a condition similar to that of the body that produced the field. It does not heat them: it electrifies them. The process is ordinarily called induction. If one would follow mentally the mechanical conditions and changes that take place when this process of induction takes place, let him imagine the two cannon-balls suspended in a room a few feet apart, and one of them to be suddenly electrified artificially in any kind of a way, as by connecting it to a charged electrical machine for an instant. The re-action upon the ether will at once begin. The stress into which it will be thrown will be propagated outwards as a wave, with the velocity of light, and equally in every direction about it too, until the advancing wave reaches the second ball, when the absorption so reduces the stress that other parts of the field can move towards it, thus distorting it; for at the outset every part of the wave moved in a radial line. This must be the case unless the field acted intelligently instead of mechanically, and knew where it was to go beforehand. Of course no one would suppose that, but the remark is made to emphasize the necessity for the mechanical steps in order to have clear ideas of what has happened. The whole would happen in so small a fraction of a second that it would be exceedingly difficult to measure it, but the rate at which a thing is done does not necessarily modify the way of doing it.

III.—THE MAGNETIC FIELD.

The distribution of iron filings about a magnet gives one a very definite mechanical conception of the shape and properties of a magnetic field. It has before been remarked that the shape of the field depended upon the form of the magnet, and when this was altered the field changed its form. That it too represents a condition of the ether seems unquestionable. That it is produced by the arrangement of the molecules of the magnet is also certain; but that presumes that the atoms themselves are magnets, each having its own field. When these atoms are either in disorder or so arranged as to mutually cancel each other's field, there is no field observable. When they are made to all face one way, their individual fields will conspire to produce a resultant field, which will be strong in proportion to the number of such individual fields that make it up. The nature of this magnetic field is probably a kind of whirl or spiral movement in the ether between the two poles of the magnet; but, as two similar adjacent whirls or lines are mutually repulsive, they spread out into space indefinitely, and are almost always curved. The earth as a great magnet has such a field, the lines reaching from the north polar regions upwards and southwards, re-entering the earth by similar downward sweeps in the south polar regions. How far away from the earth some of them may extend no one knows, but there seems to be no reason why they should not extend as far as any ray of light. There is good reason for thinking that the other members of the solar system are magnets, especially as iron and nickel are so abundant in the sun and in the meteorites that reach us from space. If that be the case, they are all moving in each other's magnetic fields. As the movement of a conductor in a magnetic field produces an electric current in the conductor, and as what are known as earth currents, apparently due to some extra terrestrial source, are well known, their origin is accounted for. But, when there is iron in a magnetic field, the latter acts upon it so as to compel it to produce a field of its own. In other words, it makes a magnet of the iron. The process is called magnetic induction. Like the other cases, it is a two-step process. There is, first, the magnet with its molecular arrangement; second, the action of the magnet upon the surrounding ether; and, third, the re-action of the ether upon the second body, making it a magnet. The heat field heats a body, the electric field electrifies a body, the magnet field magnetizes a body; and each of these fields may exist separately or simultaneously, and each do its own characteristic work, quite independent of either of the others: so the same body may become magnetized, electrified, and heated at the same time by the same medium, acted upon by three different sources. The magnetic field is more selective in its action than either of the other two. A heat field will heat any kind of matter in it if it be solid or liquid; an electric field will electrify all bodies to some degree, but solid conducting bodies to the highest degree; while the magnetic field magnetizes only iron, nickel, and cobalt appreciably, and the two latter but to a very small extent. The point of chief importance here is the function of the field itself to produce, in a certain kind of elementary solid matter, a molecular disposition and arrangement similar to that of the body which produced the field.

IV.–THE CHEMICAL FIELD.

The phenomena attendant upon the combination of atoms into molecules, and molecules in cohering together to form larger masses, make it certain that each atom has a peculiar field, which, for a name, may be called its chemical field, within which it acts upon the ether about it, and which extends to a distance from it many times the diameter of any atom or molecule.

Chemists have concluded that there is really no distinction between what has been called chemical attraction and cohesive attraction; such, for instance, as enables a drop of water to adhere to a surface, or glue to hold wood surfaces together.

Crystals are built up of similar cohering molecules arranged in a definite order. And these molecules exist as independent bodies while in the solution before being crystallized, and consequently each molecule must have some degree of attraction for others; and this is about the same as saying that there is an ether stress about each one that depends upon its temperature, for crystallization cannot take place in a solution above a definite temperature. But one of the best evidences of a chemical field of the sort is found in the fact that a solution of a given crystallizable salt has its process easily initiated by putting in a small crystal of the same kind of a substance. Moreover, the mere presence of certain kinds of molecules among others is sufficient to bring about chemical changes which otherwise would not occur; while the catalytic body, as it is called, is not changed. This is the case with starch, which is converted into sugar by the mere presence of sulphuric acid, which undergoes no change. This is apparently inexplicable, unless it is admitted that molecules of all sorts have fields which, in one degree or another, control chemical combinations. This has been treated of at some length in the chapter on chemism. Its signification here is to point out again that the field of similar molecules is of such a sort as to compel within it an arrangement of atoms into similar molecules, and molecules into similar positions, as exhibited by crystals of any sort. It is, therefore, another example of the property of a physical field to bring about in a mass of matter within it the same kind of physical phenomena as that which induced the field.

V.–THE MECHANICAL FIELD.

A sounding body sets up air waves that travel outwards radially from it in every direction to an indefinite distance. Such periodic waves are capable of making other bodies vibrate at the same rate as the original body. When the second body has the same specific rate, absorption takes place, the amplitude of vibration increases, and the case is known as one of sympathetic vibration. When the specific rate is different from that of the recurring waves, there is more or less interference, and this case is called forced vibration. In all cases, however, the second body is made to vibrate by the sound waves that fall upon it, whether the medium be the air or any other substance, solid or liquid. And the space within which such effects are produced is the field of the first or sounding body. If one considers simply the air as the medium of the field, it will be perceived that sound waves travel in every direction in it, and to distances unlimited except by the presence of the air itself. Of course, the farther the distribution goes on the less energy there will be to any cubic inch or any other dimension, and there must be some limit where the energy is too small to affect the organs of hearing; but such a limit ought not to be considered the actual limit of sound vibrations or the field of the sounding body. There is no reason for doubting that every sound vibration of every kind and degree is distributed throughout the whole earth and its atmosphere, and more than that: as the impact of molecules in sound vibrations results in heating them to a higher temperature, increased radiation into space follows, and the consequent energy in this form must affect in some degree every particle of matter in the universe upon which it falls. It is plain how far-reaching almost every act and movement of every kind must be.

A sound vibration, being a to-and-fro movement of a mass of matter, may easily be great enough to be seen, as in the case of a tuning-fork or a piano string; and, therefore, it is treated as being mechanical as distinguished from molecular: but even where the sound vibration is too slight to be seen as an actual displacement, it can give to another body a large amount of visible motion, as when a suspended marble is held against a sounding tuning-fork, or as when a paper windmill is held over a sounding Chladni plate.

The motion of a sounding body being mechanical, the field it produces may be called the mechanical field, because the effect of it upon other bodies is similar in kind to that which produced the field. There are, therefore, five well-defined modes of physical action,—heat, electricity, magnetism, chemism, and sound,—which, in the past, have often been called physical forces, each one of which affects the medium about it, producing either a stress or a motion, or both—conditions that travel outwards into space indefinitely, and constitute as many different physical fields. They may all co-exist in the same space without interference, and each one produces upon other bodies of matter within it the same physical condition of motion, position, or arrangement as that which initiated the field itself. So the established relation deserves to be called a law better than many relations that are called laws, but are such only within rather narrow limits (as, for instance, the law of Charles and Boyle's Law), inasmuch as this law of physical fields is as universal as gravitation.

What is called gravitation might be included in this list, for every particle of matter attracts every other particle near or far; so every atom has a gravitative field as extensive as the universe, and there is no more interference between it and the other fields than there is between any of them. The chief distinction between the gravitation field and all of the others is that they are all artificially variable while gravitation is not known to be, though some phenomena indicate the possibility of it.

It follows, from the foregoing, that every object large or small is continually affecting the space about it in several different ways,—through its temperature, electric and magnetic conditions, as well as by its various movements; and it also follows that the shape of a body as well as its molecular arrangement determines whether the field shall be symmetrical or otherwise. A crystal certainly has a symmetrical field, but it cannot be turned over in the hand without affecting in some degree everything outside of it.

If it be true for certain collocations of matter that external form and molecular arrangement determine the existence of its field, it is difficult to imagine why it should not hold true for all cases,—a cell structure for instance, in which case the organization of a similar cell in adjoining space where the proper material for construction exists would only be in accordance with the physical properties of fields in general; and the phenomenon of growth would be as definitely understood as the growth of a crystal. This is not demonstrative; but it is in accordance with everything else we know, and is what would be predicted by one who knew the properties of physical fields, though he had no knowledge of cell growths.

To take one step more, yet not to go beyond the domain of physics: It is as certain as any physical fact can be that every movement of an individual—change of attitude, gesture, or expression of countenance—must produce a corresponding change in his field, and tend to bring about in others similar movements; and, even if such phenomena are not observed in every one, it is no more of an argument against the existence of the operative conditions than is the failure to perceive through the sense of feeling the sound vibrations produced by a speaker's voice, when it is certain the whole body is in a state of tremor; and the effect of sympathetic speech is more largely physical than has been supposed. Strong emotions, or the physical semblance of them by skilful actors, re-act in the same physical way. This is not saying there may not be other factors, but the purely physical ones are present and act in the way described. The term "sympathetic action" was applied to physical phenomena when it was discovered to be a mode of action quite analogous to mental phenomena between individuals in which similar mental states are induced.

Lastly, so far as mental action depends upon brain structure, any changes in the latter must produce corresponding changes in the brain field, and there must be a brain field if there be any truth in the foregoing; the conclusion is inevitable. Other similar structures must be affected in some degree by them, and whether such induced changes be able to induce similar brain changes with the accompanying mental phenomena or not must evidently depend upon the possibility of synchronous action.

This is not to be understood as asserting that such thought transference as is implied in the foregoing actually occurs. All that is asserted is that the physical conditions necessary for such transference actually exist, and one who was acquainted with the properties of physical fields would certainly predict the possibility of thought transference in certain cases.

CHAPTER XIII

On Machines.—Mechanism

The common notion of a machine is that it is an implement designed for doing this or that: as, for instance, a loom is a machine for weaving cloth or carpets; a steam-engine is a machine for driving machinery; a water-wheel, for utilizing the power of water; and so on. Some of these structures, built for specific purposes, are highly complex, and many of their parts stand in curious relation to each other, and altogether they may be able to produce results that seem but little short of intelligent action. Looms weave out beautiful fabrics with artistic designs in colors, when furnished with only the bare threads. The hair-cloth loom draws with iron fingers a single hair from a large bundle of hairs. If it fails to grasp one, another and another attempt is made until one is seized, and meanwhile the rest of the machinery waits. If it seizes more than one, as sometimes happens, it drops both and tries again, the rest of the apparatus waiting as before, exhibiting a kind of deliberativeness and consciousness of what it is about that one hardly looks for through any combination of wheels, ratchets, levers, and the like, such as make up a complex machine. Every one knows that by far the larger number of things in common use which were formerly made by hand tools are now made by machinery more rapidly and oftentimes more perfect than they could be made by hand. The parts of clocks and watches are so made; papers are printed, folded, and directed at the rate of ten thousand in an hour by one machine; grass is mown, grain is cut, threshed, and winnowed by one machine as fast as it can be driven through the field; shoes, toys, and beautiful pictures are thus made by the million, and there is no department of human effort but is dependent upon mechanism of some kind. In many cases the entire work is thus done automatically, as when pins and needles are made from the wire, sharpened, polished, counted, arranged in papers, and folded ready for the market. There is no field independent of such aids. Even music is absolutely dependent upon it, and all that is called sentiment and feeling in it are resolvable into degrees and directions of movements for the production of sounds; and there are no movements of muscles but may be duplicated by automatic mechanism. If the effects produced by mechanism to-day are not the effects wanted, it only shows that the mechanism has not been perfected, not that it cannot be done.

If one considers the almost infinite number of processes needed for the maintenance, conveniences, comforts, and tastes of what is called civilized life, it might seem as if an almost unlimited number of physical conditions would be necessary; but let such an one recall the fact that all kinds of motions are reducible to not more than three fundamental kinds,—translatory, vibratory, and rotary,—and he will be prepared to trace the most complicated movements to these elementary forms.

In the chapter on motion, only the kinds of motion were considered; but here it is proposed to point out the conditions under which motion is transferred from one place to another, and how these elementary forms are transformed into each other. For convenience, the term "mechanical motion" will be employed for all having visible magnitude, but simply on the ground of visibility, not because there is any other distinction between such motions and those of a molecular or atomic kind.

When one pushes against a paper-weight on the table and it moves in consequence, no one is surprised, for the movement is expected. If the weight were free to move and it did not move, no matter how strong the push, one would have reason to be surprised, because such a phenomenon is not in accordance with the experience of mankind. If one billiard-ball in contact with another one received a push in direction toward the latter, the latter would be moved in the same direction, and the motion of the second one would be explained by saying it was due to the push of the first upon it. Suppose there were ten or a hundred such balls in a line. If the end one was pushed towards the rest of them, they would all move, the farthest one as much as the first, as the movement imparted by push to the first would be handed on step by step to the last. If the balls were glued together at their points of contact, that would make no difference in this transfer of motion by contact; and, if there were a thousand or a million, or any other number, there would be no difference. Neither would there be any difference if the separate balls were no bigger than molecules. A rod of wood or metal is entirely made up of a great number of cohering particles, and, when a push is applied to one end, every particle is pushed as much as the end particles. If there was a row of thin rubber balls and the end one was thus pushed, the side would be flattened somewhat, and the opposite side in contact with the next adjacent ball would push against its neighbor and each be flattened, and so on, till the last one was reached, which would be pressed on one side but not on the other, and would, therefore, be like a single ball pressed upon one side. The intermediate balls would act as transferrers of pressure from one end to the other. The rubber balls so flattened by pressure will recover their form when the pressure is removed, and the same may be said of a rod of any material, the difference in this particular being only one of degree. The same process takes place when one pulls upon a rod. It is to be remembered, however, that in either case the transmission of the pull is not instantaneous for any distance, however short. Time is requisite, and hence there is a rate of propagation of such motion in all bodies, which depends upon the degree of elasticity and the density of the material; and this rate cannot be exceeded, no matter how great the initial push or pull. This rate is about sixteen thousand feet per second for steel and the most elastic woods, and is about eleven hundred feet per second for air. If one inquires what the condition is that initiates motion in any given body, it will be found to be a push or a pull, and either of them may be measured in pounds. The chief distinction between a push and a pull lies in the relative position of the moving power and the body being moved by it. In the push, the body being moved leads in the line of movement; in the pull, the moving power leads. When a locomotive goes ahead of the train, it pulls; if the train goes ahead, it pushes. A stiff rod or bar may be used for either a push or a pull, but a rope can be used only for a pull, for when pressure is applied to it longitudinally it bends at right angles to the direction of the pressure, and so fails to act in the right direction. A rod can transmit a push or pull only in the direction of its length, while a rope may rest on a pulley and the pull may act upon any other body in the same plane the pulley turns in. If a pressure of ten pounds be applied as a push at one end of a rod or bar, the whole of that pressure may be transmitted to the other end. The same may be said of the pull either with a rod or rope, but neither rod nor rope can possibly transmit and give up at the one end more than is applied at the other. For this reason, a rope hanging over a pulley will hold equal weights on its two ends. If a ten-pound weight be tied to one end, the pull transmitted will be ten pounds, which may be balanced by a pull either by weight or in any other way on the other leg of the rope. The function of a pulley is to change the direction of the pull: it does not alter its amount.

MECHANICAL MACHINES.

In the older treatises on natural philosophy, there were described several machines which were called the mechanical powers, because their principles were embodied in mechanical devices for transmitting pressure or pulls. The *lever* stood first among them. It consists of a stiff rod or bar resting upon a point of support for it called a fulcrum, and this fulcrum may be placed anywhere between the ends of the bar. The advantage or disadvantage of this machine depends upon how near the fulcrum is to the body to be moved. A stiff rod four feet long supported at its middle would be balanced if it were of uniform dimensions. If a weight of ten pounds was hung at one end, an equal weight or pull would be needed at the other end to balance it. If one weight fell one foot, it would do ten foot-pounds of work in raising the other ten pounds one foot. In any case the work done, measured in foot-pounds, will be the same at both ends of the bar or lever.

The lever changes the direction of motion or the amount of pressure, but does not change the amount of work measured in foot-pounds.

The simple *pulley* is a device for changing the direction of a pull, as seen in the apparatus for raising merchandise to higher levels in buildings; but by far the most extensive use of it is in the transfer of a continuous pull from one place to another through the agency of belts of leather or other pliable material.

This combination of pulley and belt is adaptable to many places and purposes, as well as permitting great ranges in speeds of rotation by simply making the diameters of the pulleys proportional to the differences in rotation wanted. It is the chief agency in machine-shops, factories, etc., for distributing the power to the various machines. By crossing the belt the second pulley can be made to turn in the opposite direction.

In all the ways in which it is serviceable, it is plain that it cannot deliver more of a push or a pull than is given to it any more than can a lever. There is no gain of energy or work by its use, but always some loss, because friction uses up some of the working-power in other than useful ways. The *wedge*, the *inclined plane*, and the *screw* are but simple devices for utilizing push or pull; but there are other means also employed for the same purpose; for instance, the pressure of the air or other gas, and steam. Windmills are made to turn by the pressure of the wind upon the inclined blades, and, by forcing air into pipes, an increased pressure may be transmitted for long distances and then used. The reason this method of using air is not in more general use is that when the air is compressed it heats.

The heat it loses soon if conveyed in pipes very far, and as a consequence its pressure is very much reduced, so it is not an economical thing to do. Water-wheels utilize the pressure of water, and the amount of work it can do is definite and easily calculated. If at a waterfall a hundred pounds of water falls ten feet, then it can do $100 \times 10 = 1{,}000$ foot-pounds of work; that is, it can raise $1{,}000$ pounds a foot high, and so on for any other amount. A perfect water-wheel that did not let slip by any water without its doing its work would give up practically $1{,}000$ foot-pounds. Really, the best water-wheels give but about ninety per cent of the working-power of the water. So-called water-motors are but properly constructed wheels enclosed in the pipe through which water is made to flow with considerable pressure. In the cases of air, steam, and water power there is the condition we call a push, which may be measured in pounds; and a push measured in pounds multiplied by the distance in feet through which it is maintained is the measure of work.

In each of the cases, the air, or steam, or water, as it moves on and does its work, gives up the motion it has; and the substance itself, being no longer of use, is allowed to escape as a waste product. Such bodies have been sometimes called *prime-movers.*

So far has been considered only the apparatus in

common use for transferring motion of one body to other bodies, but frequently it is important to have the *form* of the motion changed from the kind it may chance to have at the outset to one better adapted to the special end desired.

In a sewing-machine, for instance, the particular movement of the needle must be vibratory. The treadle has a similar movement, but not rapid enough; so there is arranged between them a series of movable parts, which not only *transfers* a certain amount of motion, but the latter is *transformed* into appropriate forms. The vibratory motion of the treadle is transformed into the rotary motion of the balance-wheel, this into swifter rotation of the pulley by means of a belt; then by lever and cam the needle receives its proper kind of motion, the shuttle a similar one at right angles to that of the needle, and the other moving parts such forms of motion, and rates of motion, as are needful for their special kinds of work. In a steam-engine the constant pressure of the steam is made to act upon the alternate sides of a piston, giving it a vibratory motion, which must be transformed for most purposes into rotary; and this is effected by means of a crank, which is, therefore, a device for transforming vibratory motion into rotary, or *vice versa*. When the driving-wheels of a locomotive are made to rotate, their adherence to the track carries the whole structure forward; that is, the rotary motion is transformed into translatory. In the stationary engine the rotary motion of the balance-wheel is transferred to a pulley by a belt, and the shafting transfers this through its whole length to other pulleys. If the reader will follow back to its antecedents any particular motion he may think of, he will see that the function of each movable part of a machine of any sort is to transfer push or pull, or transform one kind of motion into another kind. However complex a machine may be, it does no more.

It is to be noted that *what* a given thing will or may do depends altogether upon what kind or form of motion it has, not upon how much motion or energy it has. For instance, a bullet might spin on some axis on the table before one, and have great rotary velocity and energy, yet be perfectly harmless; whereas, if it had the same amount of energy with the motion translatory, it might be destructive to anything it struck.

MOLECULAR MACHINES.

If one of the functions of a machine be to transform the kind of motion it is supplied with into some other kind of motion,—translatory into rotary or vibratory, any one into either of the others,—one may be prepared to follow mechanical processes from masses of visible magnitude into molecular magnitudes, and thus note the antecedents of the new phenomena that appear.

When a gas is condensed by pressure the individual molecules have less free space to move in, and they consequently collide with each other more frequently. Being elastic, their average amplitude of vibration is increased proportionally, and a greater number of them will strike with greater velocity upon the walls of the containing vessel per second than before. Thus the temperature and the pressure of the gas are increased. We say that mechanical energy has been converted into heat energy, or sometimes simply into heat, though what has really happened has been the transformation of external translational motion into internal vibratory motion, which the elasticity and mobility of the molecules permit. When by friction or percussion a body is heated, the same thing precisely has happened: translatory motion has been transformed into vibratory, through the agency of the molecules, which have, therefore, acted as machines for transformation.

In like manner the reverse transformation may take place in several ways. When the increased vibratory motion of the molecules produces an increased pressure upon the movable head of a piston in an engine, the piston as a whole may move and do work. Also, when the molecules strike harder upon one side of a surface than upon the other side, the surface moves toward the side of less pressure, as with the radiometer; so that both engine and radiometer are machines for transforming vibratory molecular motions into translatory mechanical motion.

When the temperature of steam is raised to about 5,000° F., the amplitude of vibration is so great that the atoms can no longer cohere in the molecules, and they become separated into the gases hydrogen and oxygen; and again vibratory motion is transformed into translatory, which in gases is called free-path.

Heat is also largely derived from the chemical properties of coal, wood, oils, gas, and other substances called fuel. As the heat is derived from some antecedent condition which is not heat, it follows that the stove or furnace is a machine for transforming into heat motions those motions which constitute and are the measure of chemism.

When heat is applied in any way to the face of a thermo-pile, electricity may appear which may be made to do work in many ways. The vibratory motion disappears as such,—that is, it is annihilated,—while an electric current appears as its substitute. The thermo-pile is, therefore, a machine for the transformation of heat into electric current. If heat be a kind of molecular motion, then an electric current must be some other kind of motion!

When the armature of a dynamo is turned and an electrical current is developed, the latter is the representative of the mechanical movement of the armature. It takes more power to make it move at a given speed when it is producing a current than when it is not. The current represents the difference. It is mechanical motion that goes into the dynamo, and an electrical current comes out of it; and hence a dynamo is a machine for the transformation of mechanical into electrical motion. One is visible, the other molecular, as is the case when friction develops heat.

An ordinary static electrical machine possesses a similar function.

On the other hand, a galvanic battery transforms chemical into electrical motions; and, in every case where electricity is developed, there is some sort of apparatus which receives one kind of motion for transformation. That one kind of machine will transform mechanical motion, a second heat, a third chemical, all into the same kind of a product, helps one to see that the antecedents, which at first seem to be so unlike, are really but varieties of the same condition, namely, motion, which, when transformed by suitable machines, might be expected to appear as a similar product of each.

An electrical current always heats the conductor through which it passes. It is, therefore, an antecedent for the production of heat in the same sense as mechanical motion is an antecedent in condensation, percussion, and friction; and the conductor is the agency for the transformation into the vibratory molecular form.

So far as the production of light by electricity is concerned, whether by the incandescent or the arc system, the function of the current is to raise the temperature of the conductor to the proper degree for luminousness. The light comes from the hot molecules, not from the electricity; but here, as in the simpler case of heating the conductor, the conductor itself, whether it be a filament of carbon or the tips of the carbon rods, acts as a transformer of electrical into heat motions, and thence to ether waves.

Ether waves may be transformed in two different ways. First, by falling on molecules of matter; the latter absorb them, and are heated in consequence, which is the converse of the production of ether waves by heated molecules. Second, by their own interferences plane, elliptical, and spiral waves may be produced, which resultant waves are capable of affecting matter in different ways. One of these consequences is of so much theoretic importance it will be well to allude to it.

Given a flexible section of a spiral ether wave, no matter what its origin. If its ends were to come together, there is good reason for thinking they would close and weld together, forming a ring, which would then be practically a vortex ring. The ends of vortex rings formed in the air will do thus, so if the atoms of matter are really vortex rings, as has been supposed, the above suggests how they may originate, or how matter is created.

All the different kinds of phenomena which are generally attributed to different forces one may readily trace to these antecedents; namely, matter, ether, and motion of various forms. The condition necessary for a new phenomenon to appear is that the present forms of motion in either matter or ether needs to be transformed. Atoms and molecules, as well as large masses of them, which we call bodies of visible magnitude, act as machines for the transferrence and the transformation of motion; and one might define a machine as a *collocation of matter having for its function the transferrence or the transformation of motion, or both.* An atom and a molecule, then, are as much machines as a steam-engine or a dynamo; and every molecule in the universe, whether near or remote, is constantly receiving and transforming energy through its individual motions.

What the particular phenomenon will be in a given case depends upon the form of the motion received by the mechanism and the new form which the latter has made it to assume. As before remarked, what a given mass of matter will do depends upon the kind of motion it has.

So far nothing has been said about the relation of these mechanical principles to living things,—animals and plants; but it will be obvious to every thinking person that unless, when matter assumes the forms exhibited by such living things, it surrenders its mechanical properties and relations, then such transformations must be going on constantly in all living things. Mechanical motions, chemical re-actions, heat, and so on, ought to be expected from such complex machines as animals. Foods, as fuel, air, and water, are physical factors which imply metamorphosis; and the forms into which the factors will be changed depend upon the special mechanism provided. Hence, an animal is a complex machine for the transformation of motions of various sorts, the sum of them being what are called the phenomena of life.

The foregoing analysis shows that what have heretofore been considered as forces in nature are non-existent; that all phenomena in the different fields of physics are simply and plainly mechanical; and that an application of the laws of motion, as presented by Sir Isaac Newton, supplemented by the laws of ether action, is sufficient to account for all kinds of phenomena: and therefore the supposition of particular forces of any kind is entirely unnecessary. What have been called forces are but various forms of motion, of matter, or of the ether, each embodying energy; the particular phenomenon a given body may produce depending upon its size and the particular quality of motions it chances to have. Granting this, one may at once perceive that expressions implying higher and lower forms of force are misleading. No one is higher in dignity or importance than any other one. Let one ask the question, Which is higher, vibratory or translatory motion? and he will see the absurdity of the language.

If one will bear these principles in mind, they will be helpful in unravelling phenomena which otherwise may appear to be very puzzling. For instance, one may frequently come across the statement that one cannot get out of a machine what is not in it or put into it. Is it so? Coal is put into the furnace, and heat comes out. Mechanical motion is put into a dynamo, and electricity comes out. A current of electricity is turned into an arc lamp, and light comes out. The character of the product thus depends upon the form of the machine and its relation to some antecedent factor. The physical knowledge we have enables us in most cases to trace and understand the metamorphosis. In some cases the molecular changes are not so completely known in detail, yet the quantitative relations between what goes in and what comes out of the machine are so definite that one is warranted in asserting that no other factors are present than the one considered. In one sense the product of any machine is like its antecedent, if both be but kinds of motion, or forms of energy as some prefer to say; but if one assumes that these various forms of energy differ in any way from forms of motion, or that they have distinct individualities, then one can get out of a machine what he does not put into it. What seem to be more unlike than the mechanical movements of a steam-engine and the electricity of the dynamo? One is simplicity itself; the nature of the other, its product, has been the despair of philosophers for generations. The subject is of fundamental importance chiefly because some philosophers have evolved their schemes without duly considering these obvious relations.

However much a given phenomenon may differ in character from its known antecedents, no good reason can be assigned for thinking that, when properly analyzed, it would be found resolvable into other factors than matter, ether, and motion. Furthermore, there is no evidence that any one of the physical forms of motion is or was necessarily prior to any other. As there is no hierarchy among them, no one of them can be called primal. A linear arrangement does not properly represent their mutual relations. They are more like a closed ring of interrelations thus:—

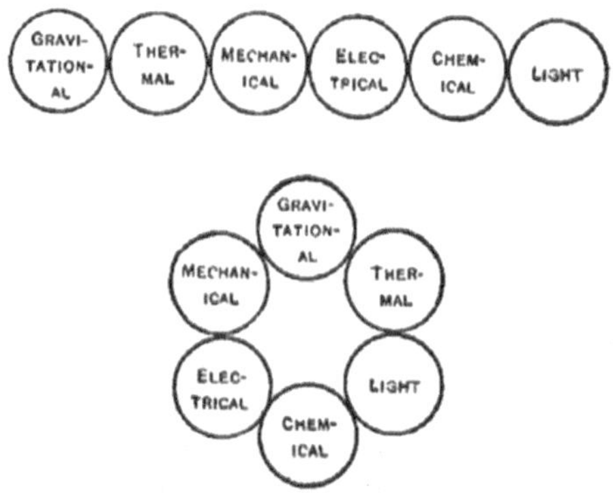

Diag. 35.—Forms of Energy.

The visible universe may be considered as a vast machine, within which motions are being exchanged by contact and by radiation. It is not the absolute amount of energy a body may have which determines whether it shall give or receive, but it is the degree it has of a given kind of energy. Thus it is the temperature of a body that determines for it whether it shall gain or lose heat in the presence of other bodies. The whole tendency is towards equalization of conditions, and for this reason some philosophers think they foresee the end of this act in the drama of the solar system. The possibility of the variety of phenomena that gives interest to existence depends upon the fact that at present matter is in an unstable condition, and, when uniformity of condition is reached, there will be an end to changing phenomena. Astronomers have figured out that in five or ten millions of years the sun will have radiated away so much of his energy that the earth will no longer be habitable. Perhaps so; but it is certain that the whole solar system is drifting in space somewhere at the rate of seven hundred millions of miles a year, and in one million of years it may reach a region in space where the present rate of loss might be greatly reduced. In that time it will have travelled three times the distance to the nearest of the fixed stars. It could hardly be where its expenditure would be greater than now. If it should drift into one of the great hydrogen regions such as are numerous in the heavens, not only would the supply of energy be renewed indefinitely, but the earth would become uninhabitable in an hour. At any rate, there is no guarantee in nature for permanent stability, supposing that stability should be attained; for simple mechanical impact between the sun and any of the millions of stars would not only annihilate the earth as such, but would so reduce to a nebulous mass the matter that now composes the solar system that the whole process of world formation would have to be gone through with again. The sudden blazing out of stars here and there in the heavens shows that similar physical processes are taking place elsewhere in the universe. Such an end is quite as probable as the refrigerating one referred to; for there is implied in the latter not only that the present conditions in the solar system will continue, but that the environment of the solar system will remain for so many millions of years what it is. The matter is not alluded to here on account of its humanitarian interest, but to point out that in either case the results will be due to purely physical conditions. What mankind would contemplate as a dreadful catastrophe would be but the interaction of huge machines, where energy was transformed on a grand scale, and no particle of matter omitted for an instant to conform to the three laws of motion.

CHAPTER XIV

Properties of Matter as Modes of Motion

In the first chapter of the book only the most obvious qualities of matter are considered, such as magnitude, density, inertia, and so on, the properties which are exhibited by masses of matter of visible magnitude and form, from which the common notions concerning its nature and possibilities have been derived. If one stops his inquiries concerning the properties of matter with these, and imagines that they are the ultimate properties, and may rightly be assumed and asserted of the individual atoms, he will be greatly in error; for it is not difficult to show that nearly every property of masses cannot be true of atoms, and that nearly if not quite all material properties of what we call matter, are derived from antecedent conditions, and are resolvable into them or into mere relations which are not inherent, and may be absent. It is, then, worth the while to study the real significance of some of the physical terms in common use, in order the better to eliminate from the mind unessential qualities when thinking of the inherent qualities of matter.

During the past ten years laboratory facilities for physical investigations have greatly aided inquirers, and added much to real knowledge in this field. Some of this knowledge is of such a character as will presently make it needful for every one to reconstruct his notions and explanations of physical phenomena, in order to prevent hopeless confusion in his own thinking.

THE STATES OF MATTER.

Under the conditions of ordinary observations matter is found in the solid, liquid, and gaseous states; the solid state being that in which the molecules cohere so strongly as not to be easily separated from each other nor from the relative positions they have assumed with reference to other molecules. Thus, a piece of granite, as the type of a solid, may have its molecules cohering to each other in certain positions, so strongly as to require a ton's weight to pull apart a section of one square inch.

The granite is made up of small crystals of quartz, mica, and feldspar, each having a definite chemical composition. The individual crystals retain their relative places for an indefinite time, and the atoms of the individual molecules retain their relative positions for a like indefinitely long time, else the crystalline structure would be lost, for crystalline structure implies definite atomic arrangement as well as molecular arrangement. So in solids the adjacent molecules are in what are called stable positions, and are not easily separated.

In a liquid there is little cohesion among the molecules, and no stable arrangement at all. The individual molecules move among each other without apparent friction, and the slightest force acting upon them makes them to turn on any axis; and there is good reason for thinking that in a liquid like water, the individual molecules are continuously rolling and tossing about with perfect freedom to move in every direction. The phenomena of diffusion exemplifies this. There is also good reason for thinking that the individual atoms in the water are continuously changing partners at a rapid rate, so if there were some means for identifying the atoms of hydrogen and oxygen in a given molecule, they might be seen presently all separated and forming temporary constituents of other molecules a relatively long remove from the first position where they were observed. When the water is frozen, that is has become a crystalline solid, this freedom of atomic change and molecular rotations is no longer recognized as a property. Molecular cohesion is now exhibited where before there was none. There are also new qualities called crystalline, hardness, density, and so on, which before this change did not belong to it. The new qualities which seem to have been developed are produced by lowering the temperature of the water, that is, reducing the amount of kinetic energy the molecules had; and by again imparting a like amount to the ice *both crystallization and cohesion are destroyed*.

A gas is a body of molecules in which the individuals are free to move in every direction unconstrained by any degree of cohesion, and where they are in frequent collisions, bounding away in new directions through distances usually many times the diameter of the molecules themselves. Thus, in air the ordinary average distance between impacts is nearly two hundred times the diameter of the individual particles which, as before stated, is in the neighborhood of one fifty-millionth of an inch. Their continuous bumping against each other and the walls of the containing vessel, produces what is called the gaseous pressure. Increasing the temperature of the gas increases the velocity of movement in the free path, and, consequently, the momentum and the pressure. It has been customary to say that heat increases the elasticity of a gas, that a gas occupies the whole space which encloses it, that a gas has a tendency to indefinite expansion, and that the properties of a gas are due to repulsive force among the molecules.

In a loose sense such expressions may be allowed, but they are not to be understood as correctly specifying the qualities of the gaseous matter. It is not repulsion that makes a ball move which has been struck by a bat, but impact; and that it should continue to move on until it strikes another body, follows from the first law of motion, as true for a molecule of a gas as for a baseball. The direction the ball takes depends upon where it is hit, as well as upon how hard it is hit; the velocity it has depends upon how hard it is hit, and there is nothing peculiar to a gaseous particle requiring the affirmation of different properties.

Some years ago improved methods of making a vacuum were adopted, by which one could reduce the amount of gas in a tube to even the hundred millionth of its ordinary amount, so that a particle might have a relatively long free path measurable in feet instead of hundred thousandths of an inch, and the phenomena of such rarefied gases were so new and surprising that it was at first conjectured that a new state of matter had been discovered, and it was called the fourth or ultra gaseous state to distinguish it from the others; but it was soon perceived that it was still only rarefied gas, and that no new qualities had been developed, and the same phenomena witnessed in the rarefied gas were present in the denser, only disguised by the greater number of molecules which took part. So what was called for a short time the fourth state has been practically abandoned.

The three states already considered are known to depend upon temperature. Thus, if ice or iron or many other solids be heated they become liquid; if heated still more they become gaseous. Some solids, like wood, when heated do not assume the liquid intermediate form, but are at once converted into a gas; but different substances have different temperatures at which they change from one form to the other. Thus, water becomes a solid at 32° Fah., and a gas at 212° Fah. Iron becomes fluid at 2,800° and gaseous at 6,000°. So far, then, it appears one might as properly speak of iron as a liquid or a gas, as of water as either, if they both may exist in the three conditions, and no temperature is specified. We do not do that, because when speaking thus ordinary temperatures are implied, but seldom or never thought of. If one had been brought up in the sun it is probable he would never have seen a solid, and if at the moon, he would know of neither liquid nor gas.

But the pressure of a gas is caused by the impact of its molecules, and is proportionate to the temperature. The law of Charles states what has been found to be true within the limits of experiment; namely, that the volume of a gas is proportionate to its absolute temperature; that is, temperature measured from an absolute zero, in which case, it is plainly to be seen, at absolute zero the *gaseous* volume would be nothing. It does not imply that the matter of the gas would be annihilated, but that the matter no longer existed in its gaseous form; the individual molecules would no longer have any free path motion, but would fall to the floor of the containing vessel, and thus remain quiescent, like so much dust. *At absolute zero there would be no gas.*

Again, in the chapter on Chemism it is shown how chemical reactions are determined by temperature, and cannot take place in the absence of heat. The late experiments of Pictet and Dewar show that as temperature is lowered chemical reactions become weaker and weaker, until some of the elements that have very strong affinities at ordinary temperatures, and so combine with energy, are incapable of combining, and appear inert at such low temperatures as can now be artificially made without great difficulty. Their experiments confirm the conclusions given on page 291; namely, that at absolute zero chemical affinity does not exist. Molecules would not only fall apart, but their individual atoms would no longer exhibit any cohesive quality; and this, it will be perceived, would render the existence of such a thing as either a liquid or a solid quite impossible, for each requires chemical action for both molecular formations and cohesion in any degree. Every kind of a structure would crumble to atoms in a literal sense. Book, tower, mountain, ocean, as well as every living organism, would completely disintegrate, and lose every characteristic property which had belonged to it. Hence, such qualities of matter as would be absolutely emptied out of it by simply reducing its temperature, cannot be considered as essential qualities. Yet when the atoms were thus deprived of what seems to us as all their useful qualities, there is reason for thinking they would still have definite form, mass, gravitation, magnetic and electric qualities which, however, by themselves could not make the mass of matter we call the earth a habitable place, nor give to life a material habitat as it now has.

It is then plainly evident that what we call solids, liquids, and gases, with all the laws that belong to each of them, are simply the relations of heat energy to groups of atoms, not the properties or laws that may be asserted of the atoms as such, and do not need to be considered by one who is inquiring for the essential endowments of matter.

There remains, therefore, an examination of the

other so called qualities to see if, perchance, they too may not, in a similar manner, be resolvable into energy relations, which, in turn, may be absent.

MOLECULAR AND ATOMIC QUALITIES.

(HARDNESS.)

Substances vary greatly in what is called hardness, and this properly serves, in many cases, to distinguish one mineral from another. The mineralogist employs a scale of ten, differing in degree from talc which is the softest, to diamond which is the hardest, and with these all other minerals are compared. But the mineralogist tells us that this scale does not represent hardness in any proportional way, because diamond is as much as ten times harder than the ruby which stands next to it in the scale; also that some diamonds are so much harder than others, that no means has yet been discovered for grinding and polishing them.

The diamond, however, is crystallized carbon, yet carbon exists in another crystalline form called graphite, or plumbago, which is soft, and may be whittled with a knife, while coke, charcoal, and lampblack are forms of precisely the same element, and these vary through the whole range in hardness. What, then, does hardness mean? Evidently it signifies the resistance offered to the separation of molecules from each other. It is the measure of their cohesion, and could have no existence in a single molecule of carbon. Furthermore, as has been pointed out, as molecular structure can have no existence at absolute zero for lack of energy needed for maintaining cohesion, *hardness cannot be a property of atoms at all.*

COLOR.

Color, either simple or compound, is exhibited by all masses of matter—for white is but a mixture of wave lengths, and no object is so black as to be invisible. Gold is yellow, copper red, lead is bluish. The petals of flowers, the feathers of birds, the gorgeous dyes of the chemist, seem to impress us with an assurance that color is a real quality of some kinds of matter, and can be affirmed of it without any qualifications. Bodies become visible either by their own luminousness as when they are hot or phosphorescent, or by the light reflected by them from some other source, as is most commonly the case. When sunlight falls upon a rose it is to be remembered that the sunlight is what we call white light; it is made up of all wave lengths which we can see. The rose petals absorb some of these waves,—the blue, the green, and the yellow, but not the red; these are rejected by the surface, and they therefore are reflected away, and testify to the selective power of the petals, *not their color,* as can be found by holding the same rose in yellow or blue light, when it will appear black, that is, will absorb all offered to it and reflect little or none. Again, when a body is self-luminous, as, for example, a piece of burning sodium which gives out a yellow light, it is to be kept in mind that the yellow rays are produced by certain vibratory rates which the atoms are compelled for the time being to make, but which the atoms will not make except on compulsion, that is, the high temperature which the heat energy gives to it, and therefore does not represent what can be called the color of the body—only an artificial state of vibration. Lastly, if all substances whatever were at absolute zero in temperature, they would be setting up no ether waves of any length, and could not effect any organ of vision, and, consequently, would not only show no color, but would be absolutely invisible. *Hence color cannot be affirmed of atoms.*

IMPENETRABILITY.

It has seemed to nearly every one who has given thought to the subject that what is called impenetrability must be a fundamental property of matter; that it was axiomatic, if anything could be, that two masses of matter could not occupy the same space at the same time. This has been believed to be true, not because it was demonstrable, but because it seemed to be reasonable. Maxwell, however, calls it a vulgar opinion. He further takes the pains to say, If hydrogen and oxygen combine to form water, we have no experimental evidence that the molecule of oxygen is not in the very same place with the two molecules of hydrogen.[1]

When it is possible to make one, a mechanical model is often of great assistance in helping one to conceive of conditions which are more or less difficult to describe in mere words; and a mechanical model that embodies this possibility of the coexistence of two atoms in precisely the same space may easily be made. Roll up a length of wire into a loose helix or spiral of any convenient length—say two feet long. Cut it in two parts of equal length, and bend the ends of each round until they touch, and fasten them thus, so as to have two rings made of spirals of wire. Each one may be taken as representing an atom of matter somewhat similar to a vortex ring, which has been assumed as the probable form of the atoms of matter. Each has

[1]See Art. Atom. Encyc. Brit., 9th ed.

form, size, and various other qualities, but if one of them be pressed down upon the other it will be found they will make room for each other, so that *as rings they both occupy precisely the same space.* This is not given here as anything more than as, possibly, a helpful suggestion as to how a seemingly impossible condition may be true. Evidently in this case the difficulty lies in the assumption that an atom of matter is a hard, impenetrable, geometrical solid, and, as Maxwell says, there is no proof that such is the fact. *Impenetrability is an unwarrantable assumption.*

ELASTICITY.

Elasticity has been assumed to be a fundamental property of atoms, and so not derivable from physical conditions underlying it; but Lord Kelvin has shown good reason for thinking that elasticity is a derived quality, for it is possible to construct models which exhibit the phenomenon in a high degree while they are in motion, and not at all while they are at rest. A number of gyroscopic disks set whirling on a circular axis, shows this in a remarkable way, and suggests on inspection, that something like such an arrangement may be the complete explanation of the quality as exhibited by atoms. A vortex ring may be considered as a large number of revolving disks on a circular axis, which will give to the ring not only rigidity, but stability of form, any departure from which will be resisted by the mechanical structure, and it will return to its original form after the deforming stress has ceased, with a rate depending upon the rate of rotation of the constituent parts of the ring. On page 47 reference is made to the behavior of a rotating disk, and how it simulates this property of elasticity. The common gyroscope may be cited as exhibiting it in a manner that depends upon the way in which the disk is mechanically mounted.

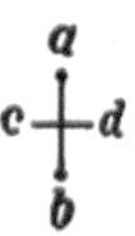

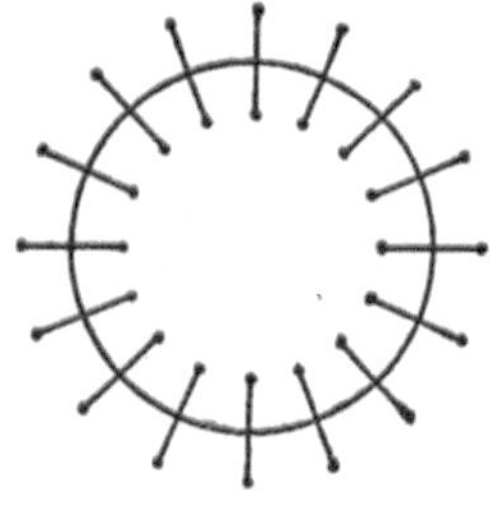

DIAG. 36.

An ordinary whirling disk can be freely moved only in its plane of rotation, or planes parallel to that. Any attempt to change the angle of the axis is mechanically resisted. This may be understood by reference to diagram where *a b* is a disk capable of rotation on axis *c d*. While rotating, it can move freely in the plane *a b*, but any attempt to tip the axis in any direction will be resisted by it. Imagine, then, a large number of similar disks, mounted on a circular axis, as in diagram 36, each one rotating. It is plain that any attempt to tip the ring in one direction or the other, or to change the form of the ring itself, supposing it to be flexible, will necessarily change the plane of some of the revolving disks, and will be resisted as a whole, for the same reason that one of its parts will do the same. It will be seen that if all these disks be rotating in the same direction the movements will be like those of a vortex ring. If additional disks could be inserted upon the axis, so as to form a continuous body quite round the circular axis, it would constitute a ring; and if the proper rotation were set up, it would possess all the qualities of a vortex ring, and elasticity would be a prominent quality, as stated on page 45. It is no longer necessary, if it ever was, to assume elasticity as a fiat quality, imposed upon atoms which might have existed without it; *for the laws of motion, acting in a properly constructed mechanism, are quite sufficient to produce it.*

MAGNETISM.

On page 246 the statement is made that magnetic phenomena have led to the belief that all atoms of all kinds of matter are magnetic, and are only obscured in ordinary matter by the molecular arrangements which tend to neutralize the magnetic fields of the individual atoms. Attention is again invited to the diagrams on page 245, with the accompanying description, in order to freshly bring to mind how vortex motion necessarily produces what is called polarity—the two sides of the ring have different qualities. On one side the movements are all inwards, on the other outwards, and from these the phenomena of apparent attraction and repulsion necessarily follow. But beyond this, once assume that individual atoms are magnets by virtue of their constitution, and that every magnet has a magnetic field infinite in extent, within which it can affect other atoms, one can see at once that every atom in creation has a magnetic hold upon every other atom, because every one is in the magnetic field of every other one. This effect is not necessarily one of attraction or repulsion tending to move one mass towards or away from another; but, on the other hand, it tends to rotate each on an axis so both shall face the same way. So long as there is no change in position or in *form* of such a magnet, the magnetic field will be uniform; but if the form be changed in any way the whole field has to change in conformity with it, as described on page 302, and such vibrations as constitute the heat of an atom are really the change in form of the atom, and this, therefore, changes necessarily the whole magnetic field of the vibrating body. These changes in the field, which originate in this way, are what are called ether waves. When the waves are produced slowly, by an alternating dynamo current, or more swiftly by some of the methods so ingeniously devised by Hertz and Tesla, they are called electro-magnetic waves. When produced so swiftly as to have a wave length only the one thirty-thousandth of an inch, they have been called heat waves; and when the waves are so short as to be capable of affecting the retina of the eye, they are called light waves, though there is no distinction between any of them except in their length. The vibrations of the atomic magnet are rapid because it is small; the waves it produces are changes in its magnetic field in the ether, so one may trace back in this manner the phenomena of light, of heat, and electricity, to the mechanical structure of atoms; and it is mechanically intelligible too, and, like the preceding accounts of properties, it appears *that magnetic and electric qualities are due to the peculiar kinds of motion embodied in the atoms,* and cannot be considered as particular endowments of a something called matter, which it might have been without.

INERTIA.

The inertness of matter has been touched upon on page 84, and here may be added the consideration of what interpretation could be put upon such phenomena as are exhibited by such a device as is represented by diagram 36, supposing it were enclosed in a box so one could not see the mechanism? The box enclosing it would exhibit a new quality of the nature of inertia, by virtue of the motions within it, which it would lose as the friction diminished the rate of motion, and when this stopped altogether the property would no longer be present. *Hence, inertia, too, must be looked upon as probably due to motion.*

MASS.

Mass, as a property of matter, is generally defined as the amount of matter considered, and is measured by what is called acceleration, that is, the velocity it acquires in a second when acted on by a constant force or push. Amount of matter is a very indefinite expression, but is often convenient, and seldom misleading when one is considering a given weight of a substance.

One may speak of a pound of iron or of hydrogen as a mass of iron or of hydrogen, meaning a definite weight made up of a very large number of molecules of one or the other element; but if one will think of the atoms of these, and endeavor to form an idea of what can be the physical meaning of mass, when applied to one of them, he will at once see that the term carries with it no conception whatever of the physical difference between atoms of different kinds. An atom of iron is said to contain fifty-six times the mass of an atom of hydrogen, while an atom of gold has a hundred and ninety-six times the mass of the hydrogen atom, and all the elements differ in mass in the ratio of their atomic weights. Can any one suppose for an instant, that an atom of gold is a hundred and ninety-six times larger than an atom of hydrogen? There is some evidence that atoms differ somewhat in magnitude from each other, but none of any such difference as is represented by their atomic weights. Furthermore, this would imply that atoms were blocks of some primeval stuff of uniform quality, and that atoms of a given element were but uniform volumes of it; and it hardly needs to be said that such a view is negatived by all we know, for the properties of the various elements do not vary simply with their weights, as would be the case if they were thus constituted. Hence, mass as applied to atoms cannot be thus conceived. It is possible to form a conception of the physical meaning of mass as applied to atoms or molecules, by recalling the phenomenon of rigidity in position, which is the outcome of rotations, as described on pages 47 and 414, for the amount of effort needed to move such a rotating body depends not simply upon the amount of rotating material, but its velocity of rotation; so a small amount of material with a high speed may offer as great a resistance to movement from its position as another much larger amount of material with corresponding slower rate, but otherwise the two would necessarily have great differences in their other properties; thus their rates of vibration would be very different, because their degrees of elasticity would be different.

One may then assume that such differences between *the atoms of the elements as are called their masses, are due to the relative rates of rotation.* This, of course, on the fundamental assumption that the atoms themselves are vortex rings such as we have argued as being highly probable.

GRAVITY.

There now remains to be considered one more general property of atoms and all combinations of them; namely, their gravitative property. If one be content to say that not enough is known about it to warrant even a tentative opinion, and, therefore, refuses to draw any inferences from what is known as to what gravitative property is, or is like, one need to have no quarrel with such an one; but if, on the other hand, one is interested in fundamental questions, and thinks that whatever be the truth about gravitation or any other unsolved problem, when it is known, it will be seen to be in harmony with every other physical truth, and will, therefore, be a consistent part of the body of physical knowledge which we now possess, such an one will perceive that with the banishment of the old notions concerning the structure of matter, with its endowments of sundry properties which might have been otherwise, or that the matter we know might have had entirely different properties, must also go the notion of quality endowments in any such sense as was formerly held. He will also have good reason for holding it altogether probable, that if the other properties of matter are reducible to modes of motion, so the last one in the list will be found to be reducible to the same factor. If the others have been interpreted thus, one after another yielding as molecular phenomena became better understood, he will conclude that if the problem of gravitation has not been solved, as the others have been, it is not because it is insoluble in itself, but because it is inherently more difficult, or has not received the degree of attention that has been given to other problems since conservation has been discovered and forms a part of every discussion.

One thing seems certain, if the vortex-ring theory of matter be true, or anything like it, then gravity must follow from the structure; for in the absence of any evidence of the existence of gravitation in the ether, no one is at liberty to postulate it there for the sake of finding it in the atoms. It must be looked for as due to the particular kind of motion that constitutes the atom, and is constant because that motion is constant.

In the chapter on gravitation is given a mechanical conception of gravitative conditions which, whatever may be its inadequacy, is consistent with other physical knowledge. Faraday, as is well known, made several efforts to discover some relation between gravitation and electricity, but only negative results were reached. He was not discouraged by his lack of success, and had planned still other experiments, which he was not able to finish. He always worked on some hypothesis almost always radically different from the hypotheses of his scientific contemporaries, and time has vindicated his rather than theirs; and that there must be some physical relation between the two classes of phenomena was one of his, and so it seems to-day; for if gravity be due to the form of motion in the atom, and if an electric current in a circuit represents a real vortex ring, having the conductor for its core, then it seems likely there is some gravitative effect between such current and the earth; but it may be so slight with a single circuit as to not be detectable with present means, and the mutual gravitative effects between two such circuits would be obscured by their electro-magnetic effects.

Lastly, if the atom itself be a vortex ring, as explained in the chapter on the ether, it follows that in the absence of such form of motion there would be no atom—no matter, though the substance out of which the ring was constituted would exist, but without any of the characteristics that we assign to matter in any of its forms. If one chooses to call a common smoke-ring *alpha*, evidently when the ring is dissipated there is no more ring, there is no alpha, it has been annihilated as a ring; and in like manner, one may understand that what constitutes an atom is not so much the substance it is composed of as the motion involved in it. Such *an atom is a particular form of motion of the ether in the ether*, in the same sense as what is called light is a form of motion of the ether in the ether. One is an undulation, the other a vortex. One we call an ether wave, the other we call matter: both involve energy, and both have properties. Thus, one after another of the properties of matter are found to be resolvable into ether motions, ether being the primal substance, and matter only one of its manifestations.

Such a conception of matter as is here presented, resolving as it does all its physical properties, even itself, into modes of motion of the ether in the ether, is not simply a new conception of matter, it is rather a revolution in fundamental conceptions, and if trustworthy, necessitates an abandonment of nearly every notion concerning them which men have entertained when thinking and discoursing upon the subject. The mystery of phenomena is not lessened but made greater by the discovery that everything which affects our senses in every degree is finally resolvable into a substance having physical properties so utterly unlike the properties of what we call matter, that it is a misuse of terms to call it matter.

No one in the past has been able to forecast its properties. The necessity for such a medium has not been felt by many philosophers, and though there has been some expectation on the part of a few, any new step has been a source of surprise. For instance, when Hertz succeeded in producing electro-magnetic waves in the ether only two or three feet long, it was heralded as being a demonstration of the existence of the ether, implying that all the phenomena of induction and electro-magnetic waves developed by machines vibrating up to four thousand per second in telephonic apparatus, could have some other interpretation. The point here emphasized is that the properties of the ether and their relations to such physical phenomena as have been the subjects of research are so little known, that no one has yet ventured to embody them in an all embracing philosophy, so as to deduce apparent phenomena from them.

The significance of this will be apparent when one recalls the various attempts of materialistic philosophers to explain all sorts of phenomena as due to matter and its properties. Some of them have been ignorant of the existence of the ether; others have grouped matter and ether together and called both matter, and considered both as subject to the same laws as are found to hold true for matter as defined in this book. When it is apparent that such physical views are radically unsound, that one cannot reason from our perceptual matter to imperceptual ether,—for it is true that there are no known nerves that respond directly to ether action,—it will also be apparent that any scheme of things that ignores this knowledge or fails to make proper distinctions here cannot be entitled to respectful consideration. Indeed, such physical materialism is less rational than ever, for it ignores much knowledge now in our possession which is as certain as any we possess, and it ignores the trend of all the physical knowledge we have; for it cannot be denied that the advance in knowledge which has been so marked during the past half century has been in the discovery of the simplicity of relations, rather than towards ultimate explanation. It may truly be said that, in a philosophical sense, nothing has been explained. Familiarity with constant phenomenal relations induces in us expectations of certain happenings, and presently they seem obvious. The car moves because the engine pulls it; the engine moves because the steam pushes it; the steam pushes because the heat pushes it; and the heat pushes because—it is the nature of heat to do work. In that way, every physical phenomenon runs at last into an inexplicable, into an ether question; and the necessity for it follows from nothing we know or can assume. No one may assume for an instant that the possibilities of ether phenomena are limited by such interactions as have hitherto found expression in treatises on physics. Indeed, there is already a body of evidence which cannot safely be ignored, that physical phenomena sometimes take place when all the ordinary physical antecedents are absent, when bodies move without touch or electric or magnetic agencies,—movements which are orderly, and more or less subject to volition.

In addition to this is still other evidence of competent critical observers that the subject-matter of thought is directly transferable from one mind to another. Such things are now well vouched for, and those who have not chanced to be a witness have no *a priori* right from physics or philosophy to deny such statements. Such facts do not in any way invalidate physical laws, nor make it needful to modify present statements concerning energy. Physical laws are not compulsory; they *rule* nothing; they are but statements of our more or less uniform experience. If these things be true, they are of more importance to philosophy than the whole body of physical knowledge we now have, and of vast importance to humanity; for it gives to religion corroborative testimony of the real existence of possibilities for which it has always contended. The antecedent improbabilities of such occurrences as have been called miracles, which were very great because they were plainly incompatible with the commonly held theory of matter and its forces, have been removed, and their antecedent probabilities greatly strengthened by this new knowledge; and religion will soon be able to be aggressive with a new weapon.

CHAPTER XV

Implications of Physical Phenomena

A physical phenomenon is a phenomenon which involves energy. Every change of condition in matter is brought about by the action of energy upon it in one way or another. It may be gravitative energy or heat or light or electric or any other; but every physical change has a physical antecedent as well as a physical consequent, and the explanation of any given phenomenon consists in pointing out the precise antecedents that brought it about. There is a common saying that like causes produce like effects, but this is far from being true in the popular sense. If it were true the development of science would not be the difficult and painfully slow process it has proved to be. Electricity may be produced by turning a crank, by dissolving a metal, by twisting a wire, by splitting a crystal, and in others ways. The product is the same, but the antecedents are so different that no one can tell by examining the product how it was produced. If it became important to know what caused the electrical phenomenon, it would not be sufficient to know that electricity could be produced in these different ways; one would need to know the specific apparatus employed. The more complicated the phenomenon the more difficulty there is in unravelling it.

So far as experiment and experience have led us, the antecedents of every physical phenomenon are themselves physical, and more than that, all reactions are quantitative, that is, the product is proportional to the antecedent, and this is sometimes embodied in what is called the doctrine of the Conservation of Energy which every one knows about.

The exchange relations between the different forms of energy,—mechanical, thermal, chemical, electrical, etc., which are so well-known, being quantitative, are therefore mathematical. They have therefore become a corporate part of the body of knowledge, and are no longer subject to any questions as to their validity under any circumstances whatever. One who should challenge them would no more be deserving of attention than if he should offer to prove he could square a circle.

The fundamental postulates of physical science are binding upon the one who understands them, for the same reason that the multiplication table is. There are no contingencies and no possibilities of hedging. If any one of them could be overthrown the whole body of science would go with it. This is said because there are not a few who appear to think that what is called physical science may not be so certain as its advocates think, and that there may be factors which have not yet been reckoned with that may quite transform the whole scheme. Science is a consistent body of relations, not simply a classified body of facts. These relations have been discovered by experiment, not by deduction.

Some of them are the following:—

1. Physical changes affect only the condition of matter, not its quantity. One cannot create or annihilate it, nor can one element be changed into another.

2. Every atom is continually exchanging energy with every other atom, the rate of the exchange depending upon their difference in temperature.

3. The different forms of energy are transformable into each other, but the quantity of energy is not altered by the transformation.

4. Complex organic molecules differ from simpler inorganic molecules in possessing more energy. The differences in this respect are definite, may be measured in foot-pounds, and are practically enormous.

5. Every physical change has a physical antecedent, is therefore mechanical, and is conditioned by the laws of energy.

These principles are the outcome of modern investigation, the evidence for them is overwhelming, and a working knowledge of them needs to be a part of the mental equipment of every investigator, especially of the one who takes it as his province to explain phenomena.

Science is strong here if it is anywhere; and any description of any event, any explanation of a genuine phenomenon that practically ignores these, cannot be true, and can have no claim to consideration.

Before any explanation is needed there is always the advisibility of ascertaining that the alleged event really happened, and whatever is not professedly miraculous must not be in discordance with the bast knowledge we have.

With the above principles in hand one is prepared to fairly judge as to whether a given statement is credible or not. It is not necessary, as some seem to suppose, that one should be able to explain a phenomenon if he rejects the explanation of another one, or to assert with emphasis whether a thing is possible, probable, or impossible.

In La Fontaine's fable the philosophers were at the theatre witnessing a play in which Phœbus rose in the air and disappeared overhead. They undertook to explain the phenomenon. One says Phœbus has an occult quality which carries him up. Another says he is composed of certain numbers that make him move upward. Another says Phœbus has a longing for the top of the theatre, and is not easy till he gets there. Still another says Phœbus has not a natural tendency to fly, but he prefers flying to leaving the top of the theatre empty. Lastly, a more modern philosopher thinks that Phœbus goes up because he is pulled up by a weight that goes down behind the scenes. The last is an explanation. From a physical standpoint the others are not simply inadequate explanations, they are absolute nonsense. They make the antecedents of a phenomenon involving energy, factors that have no more relation to energy than has moonshine to metaphysics. Yet there has been a large number of men in all ages, men able in many ways too, who have ventured to explain phenomena in such a *non-sequitur* way, and who have spurned the mechanical philosopher and his explanations.

In that class of phenomena called spiritualistic there is a large body of reputed physical phenomena, vouched for by large numbers of witnesses, such as the movements of furniture, chairs, tables, books, pianos, etc., the playing upon musical instruments, guitars, accordions, pianos, the appearance of lights, of faces, of full forms clothed, of conversations with materialized spirits, and so on, in great variety.

I suppose no one doubts that to move a body of any

magnitude requires the expenditure of energy, and to do a definite amount of work requires always the same amount of energy, yet I suspect there are many persons who give credence to statements of occurrences which practically deny the above proposition, thinking it to be probable that spiritual agencies may have control of powers that mankind knows nothing about. This may be true enough, but the question is not as to what this or that agency can do, but whether if spirits do a certain kind of work it takes less energy than if a man should do the same thing.

Whenever a weight or a resistance and a velocity are given, it is always possible to compute the energy spent to produce or maintain it. Let us study a case or two. In olden times it was related that one of the prophets was carried through the air by the hair of his head from Babylon to Jerusalem. In later times it was said that Mrs. Guppy was similarly transported from Edinburgh to London. The distance is about 400 miles, and if I remember rightly she made the transit through the air in less than one hour. This makes the velocity to be about seven miles a minute or 600 feet per second, which is three times faster than the highest tornado velocity. The resistance offered by the air to the movement of bodies in it is very well known. Pressure in hurricanes has been observed as high as 90 pounds per square foot, and as the pressure increases with the square of the velocity, it follows that at 600 feet per second the pressure per square foot would be about 800 pounds; and if the exposed surface of Mrs. Guppy was no more than six square feet, the total air pressure must have been not less than 4,800 pounds. Now, the energy of this is found by multiplying the pressure by the velocity per second.

$$4,800 \times 600 = 2,880,000 \text{ foot-pounds,}$$

and as a horse-power is equal to 550 foot-pounds per second, it follows that it took not less than

$$\frac{2,880,000}{550} = 5,236 \text{ horse-power}$$

to move Mrs. Guppy in that way at that rate.

It was reported when Madam Blavatsky was living that she was in the habit of receiving letters from distant correspondents, brought to her by some occult agency and dropped upon her table. These letters were said to have been written only a few minutes before by persons living in the most distant parts of the earth.

It takes but a little figuring to discover the amount of energy necessary to do a work of this kind. Thus, let the distance be 10,000 miles, the time ten minutes. The pressure per square foot due to such a velocity in the air will be 17,000,000 pounds, or 118,000 pounds per square inch. Assume but one square inch as the area exposed to such a pressure, then the energy needed to transport it with the speed of 16.6 miles per second, will be

$$\frac{118,000 \times 5,280 \times 16.6}{550.} = 18,000,000 \text{ horse-power.}$$

Unless such packages were protected by occult agencies also, they would be burned up before they had gone the first mile of their journey.

The popular idea is that at death the spirit leaves the body, but that it may, and often does, remain in the locality, and is frequently in the presence of its friends, unperceived by them, though occasionally they may be seen and communed with through the agency of certain preternaturally gifted persons called mediums.

This proposition has so many physical data, and involves so many physical implications, it will be worth the while to look squarely at some of them.

1. A spirit is supposed to be a conscious entity dissociated from matter, having ability to move at will and to be more or less interested in what is going on in the world, and capable of giving information on matters remote from observation or the knowledge of men. Suppose then such an entity, a disembodied spirit, without a corporeal body, but anxious to be in

the neighborhood of its former friends. Seeing that it now has, according to this view, no longer a hold upon matter, it has ceased to be in any way affected by gravity and inertia, for these are attributes of matter. Now the earth has a variety of motions in space; it turns on its axis, so that a point on the equator is moving at the rate of a thousand miles an hour. It revolves about the sun at the rate of nearly seventy thousand miles an hour, and with the sun and the rest of the bodies that make up the solar system it is drifting in space at the speed of sixty thousand miles an hour or more, so that the actual line drawn in space by any point upon the earth is a highly complex curve drawn at the rate of upwards of a hundred and twenty-five thousand miles in an hour. Now, any object whatever keeping up with the earth, but without the help of gravity, must maintain the velocity in space of not less than a hundred and twenty-five thousand miles an hour, and that is not all, as the movement is not in a straight line, any such object wishing to keep in a particular locality, say a room, would have to be on the alert constantly, for the earth wabbles for numerous reasons and what seems to us, who have bodies held by gravitation to the earth, as so quiet and smooth running that we are never conscious of the motion for an instant, is so simply because gravity takes care of us. Once surrender that and undertake to depend upon some supposed private source of energy, and one would instantly discover he had an engineering problem of a high degree of complexity. If one assumes, as some have done, that such spirit is composed of, or associated with, some sort of matter, and that navigation is accomplished by an act of the will, it will not change the foregoing factors in the problem at all.

2. Suppose, as some have done, that disembodied spirits lose their hold upon matter, and that they do not remain at the earth. Then, if they remain at the point where separation from the body took place, in an hour the earth will have moved forward one hundred and twenty-five thousand miles. But over the earth there is certainly a death every minute all the time, and such are left in the rear by the earth never to return to them, for the movement of the earth is not a circuit, but an apparently endless drift. Think of the dead of the earth for the thousands of years since man has lived upon it! On this view, the spirits might be seen like the tail of a comet reaching backwards for millions on millions of miles,—the trail of the dead.

In any view, time and space and energy cannot be ignored or ruled out.

At *séances* the reported phenomena are mostly of

a physical sort, the trance of the medium being a physico-mental phenomenon. The phenomenon of sound implies the expenditure of energy, it is a vibratory motion of the air or other elastic body, and in order to produce it some antecedent force must be spent; it may be produced by mechanical means, or heat, or electricity, or by the muscles. Its production does not imply any specific method any more than articulate speech implies a person, as Faber's talking-machine and the phonograph prove.

Let us consider some of the more subtle phenomena that are reported. First, as to so-called conditions. One of the primal ones of these for such phenomena as the movements of bodies and materializations, is said to be darkness. This is of so much importance that it must be fully attended to. To one who has not paid any attention to what has been done in molecular science within the past fifteen or twenty years, the phenomena of light may and probably do seem to be due to an unique agency, as much as heat or electricity; and therefore he looks upon light as he looks upon the others in the hierarchy of the physical sciences, and expects that in its absence a potent agency or kind of energy is lacking. That this idea and conclusion is all wrong will be apparent when it is recognized that *what we call* light is a particular sensation in the eye, and that to produce the sensation *there is no one antecedent that is essential.* Press the eye with the finger in the darkest night and one will see a ring of light with great distinctness. An electric shock, a bump upon the head, will also give one the sensation of light, and in the absence of other aids to a judgment no one could tell what was the antecedent of a given light sensation.

Radiations from a luminous body, and reflections from a non-luminous one, were not long ago thought to consist of three different kinds of rays,—heat, light, and actinic rays. It has been discovered that there is no such distinction in fact. What a ray will do depends upon what it falls upon. The same ray that falls upon the eye and produces the sensation of light, would heat another body, or do photographic work. The only difference in rays is in their longer or shorter wave lengths, and the energy of a wave does not depend upon its length. From this it follows that there is no such thing as light as distinguished among forces or forms of energy. *Light is a sensation,* and in the absence of eyes no such distinction could possibly be discovered. Light, then, as a particular kind of agency takes no part in phenomena outside of the eye. The eye of man is adapted to respond to certain wave lengths, the eyes of other animals are adapted to respond to other wave lengths; and if our eyes were adapted to perceive all wave lengths the whole universe would be always light about us, every object, whatever its temperature, could always be seen as easily as we now see when the sun shines.

These facts make it quite impossible for a physicist to understand why darkness should be an essential condition for the occurrence of such phenomena as are described. Again, every ray of light when traced back leads to a vibrating molecule or atom. Indeed, light or ether waves in general all imply vibrating atoms or molecules; and what is called spectrum analysis is but a development of this fundamental principle, and not only the kind of matter, but its physical condition is revealed. If Moses had had a spectroscope when he saw the burning bush it might have told him the nature of that conflagration.

So when luminous forms appear at a dark *séance*, there is first the ether waves of such length as to affect the eye; these traced to their source must arise from vibrating molecules, that is, matter expending energy in the production of ether waves; for no matter ever shines without some source of energy.

If the matter that gives out the light be ordinary matter, there is no difficulty in understanding it; for matter can be made to shine in several ways,—by impact, by high temperature, by electric vibrations, by chemical reactions; and no one could tell from the simple fact that the matter shone, what the origin was. But it is said that these forms that are seen and thus affect the eye, that are touched and thus affect the sense of touch, that are warm and thus testify to vibrating molecules, that speak and appeal to the ear through air vibrations, are *materializations*; meaning by that that the body with its various organs and their functions is built up *de novo* out of material at hand, as Adam was said to be made of the dust of the ground, and as the lion that pawed to free its hinder parts from the soil out of which it thus grew. What are the materials that make up a human body? Ultimately there are carbon, hydrogen, oxygen, nitrogen, iron, phosphorous, sulphur, potassium, sodium, and several other ingredients of less importance. From a hundred to a hundred and fifty or more pounds of these are needed for one full-grown person.

Many of the materializations that have been described, from Samuel the prophet to Katie King, have appeared to be veritable specimens of humanity even to avoirdupois and all that is implied in that. If the matter of such bodies was a creation and not a collocation, then one of the fundamental principles of physics is simply not true; for matter can be created and annihilated by any spirit that knows how to find a suitable medium. If the material is gathered from the environment—and this sometimes is asserted—then the difficulty is nearly as great.

One must take notice of the difference there is between inorganic or relatively simple chemical compounds and those that make up the bodies of living things,—the bones, the tissues, the muscles, the nerves, the brain, the blood. For building up a single pound of such tissue as muscle or of fat requires the expenditure of energy represented by about sixteen million foot-pounds; and as in such a body as we are supposing there could hardly be less than twenty-five or thirty to be so reckoned, it follows that not less than four hundred million foot-pounds of energy is necessary, a quantity equal to upwards of twelve thousand horsepower, if done in a minute; and if done in half a minute, then twice that quantity. I cannot but wonder if those who think they have witnessed such phenomena could have been conscious of the stupendous amount of energy which was being evolved before their eyes. Then dematerialization involves the annihilation of the same amount; for it is to be remembered that organic matter differs from inorganic in the amount of energy absorbed. There has been either the creation and annihilation of matter or the creation and annihilation of an enormous amount of energy, without antecedents and with no residuals. This is not saying that such events have not taken place, it only points out the factors of energy which are implied if they do happen.

One who is unaware of such implications and phenomena may easily suppose the most improbable things can take place. Those who are aware of such implications cannot hear of such events without instantly perceiving how almost infinitely improbable they are.

Reports of such phenomena have never come from any man who understood the relations of phenomena.

Scientific men have been often told of their incompetency to investigate so-called psychical phenomena; but if the latter involve physical phenomena, then who else can properly investigate them?

This paper is not to be understood as implying that there is no relation between the living and the dead, for the writer does not believe that doctrine; instead of that he thinks we are very near to a discovery of a physical basis for immortality that will transform most all our thinking. If spiritual communication is not accompanied with physical phenomena in the alleged way, it does not follow that it may not happen in other ways that do not do such violence to our fundamental knowledge as most of the reported cases do. The universe is large, not much of it has been explored. We live and move and have our being in an environment about which our knowledge is most meagre; but our knowledge of energy we get not only from the earth, but from the sun and most distant stars and nebulæ, and it is not probable that any contribution whatever will materially modify our present knowledge of it.

Thus far I have considered what is always implied when physical phenomena are considered, especially with reference to the antecedents; for instance, when a steam-engine is run it implies the consumption of fuel, which in turn implies molecular structure, and a definite amount of energy in what is called its chemical form. That energy is not created or destroyed by any physical process, and, therefore, every exhibition of energy, no matter where or when, is to be explained solely by reference to the laws of energy which are now so well known as to have passed out of the region of conjecture or hypothesis. If there be any knowledge which man possesses, which for certainty and accuracy compares with mathematical knowledge, it is the knowledge of physical relations. I traced out a few cases in which the alleged phenomena were of such a physical sort as to be easily handled, and showed how one must look at their antecedents. That such phenomena did take place was not denied. It was simply asserted that when they did happen one must reckon with the implications, unless he was prepared to affirm that physical phenomena might happen when physical laws are ignored and quite counted out. There are yet some further implications it is well to consider. They have to do with the objective structure and qualities of the spiritual beings that are supposed to bring about the phenomena we are considering, such as moving objects, playing upon musical instruments, writing upon slates, and so on.

As such beings are always addressed as if they were visible personages, possessing the same organs of hearing, seeing, and so on, as are possessed by individuals still having a material body; and as the replies to questions never contradict such assumptions, but, on the contrary, are confirmatory of such assumptions, it follows that one may properly consider what really is implied in the assumption that spirits have eyes and ears, because they can see and hear. When I say *I see*, I assert not only the existence of what we call light, but the existence of an organ called the eye, the structure of which is adapted to be acted upon by what we call light. Light is, as we all know, a wave-motion in the ether. It travels at the great velocity of a hundred and eighty-six thousand miles in a second, and the waves are in the neighborhood of only the one fifty-thousandth of an inch long. The eye is the only structure in the body that can perceive these waves. It is a kind of camera, and photographic work goes on in the retina very much as it does in the process of photography. Then, there is the optic nerve, which is an essential part of the apparatus, and conveys to the seat of consciousness the impress of the molecular disturbances which have taken place in the eye. No one is conscious of the phenomenon of light except through the action of this complex mechanism. Therefore, when one says he *sees*, he means that a particular kind of disturbance has taken place in a particular physiological structure. The term sight is never used in a different sense from this, except when it is avowedly used figuratively. In the absence of ether waves there could no more be what we call sight than if there were no eyes; both are essential.

When, then, it is said or admitted that a spirit *sees*, not in a figurative sense, but in the sense in which we all use the term, it is implied that a spirit has eyes, a physiological structure, acted upon by ether waves, and the nervous system behind that. It has what *we* call eyes. It will not do at all to say that such spirit has an equivalent sense, for whatever that might be it would certainly not be *sight*. One may get a very accurate knowledge of the presence of another person by the voice, or by the sense of touch, but it would be a culpable misuse of language to say of such person that he was *seen*. Sound can no more affect the eyes than light can affect the ears. This, then, is the same as saying that a spirit has a physical structure for seeing similarly constituted to that in man, and, indeed, in all organizations that *see*.

When I say *I hear,* I mean that air vibrations have affected my organs of hearing, the ears with the nervous structure between the ear and the seat of consciousness. There is implied in the statement not only that sound vibrations of a definite sort have been produced and are acting, but that they are acting upon a certain physiological structure adapted to be affected by gaseous vibrations. Vibrations in the ether cannot affect the organ of hearing. The media are radically different, and cannot be used as substitutes for each other; and it is therefore wrong to say *I hear,* unless what I perceive reaches my consciousness through the physiological mechanism called the auditory apparatus. In a figurative sense one may say he hears as he may say he sees.

> "Lo, the poor Indian! whose untutored mind
> Sees God in clouds, or hears him in the wind."

But real seeing and real hearing imply certain distinct organs adapted to different physical conditions.

One cannot, by talking, affect one's eyes; nor will light waves, as such, affect one's ears.

Suppose, then, in a *séance*, when a spirit is addressed thus: Will the spirit please rap upon the table? and the answer comes at once,—a rap distinctly heard. The question was an oral one, and was produced by physical means, regular sound vibrations, and can be heard by such beings as are possessed of the proper organs to be acted upon by air vibrations, that is, ears; and by ears I *mean* ears, not substitutes of any sort. What we call *speech* is absolutely impossible in a vacuum, as much as is sound, for speech is a succession of sounds. There are numerous substitutes for speech,—signs made with the fingers or lips that do not appeal to the ear; but these are not speech. If, then, spirits *hear*, it is because they have ears, organs that can be affected by sound vibrations in the same manner as we, the so-called living beings, can be. Moreover, do not all testify that they can and do both see and hear?

In like manner one may treat of the sense of feeling, or any other sense. All imply a molecular structure, a nervous organization, indeed, everything that goes to make up a consciousness of the external world such as is possessed by living beings governed by physical laws.

It is clear that what we call pain is immediately due to disordered nervous structure, and in the absence of nerves could never be known. This can be tested in a minute by any one, by simply pricking one's finger. Does not the destruction of the nervous tissue in any manner end the possibility of pain? Can a spirit then suffer physical pain without a nervous organization? By pain I mean what all mean by the term, the sensation which, if severe and long-continued, results fatally to the sufferer, because the nervous tissue is itself destroyed.

If some one having read so far, perhaps with impatience, should say, "All this may be as you say for living beings, incorporated in a body of flesh and blood and a nervous system, but we are not to suppose for a moment that spirits are thus constituted, and if not, then they are not to be supposed to be conditioned by such physical laws as all common matter is conditioned by. They have their own constitution, different enough from ours, and one cannot reason from our condition to theirs." To this I would reply, that if one cannot do this, if a physicist must not carry his terms and conceptions into this spiritual domain, for precisely the same reason the spiritualist must not talk about a spirit *seeing, hearing, feeling,* and so on, unless he admits he is talking loosely, and means by those terms only to symbolize his conceptions, and has to employ such terms as best convey the idea, which idea cannot be physically true. Even then it is very difficult to understand why, if the physical terminology is inappropriate, any one should at a *séance* ask such a question aloud as, If John be present will he please rap on the table; for this is *sound* addressed to an ear—both of which are purely physical things.

An Arab may not have any difficulty in imagining a genie that may be summoned by rubbing a cup, to do wonderful things, and then vanish out of relations to everything; but no one who has studied deeply into the significance of physical relations can possibly admit that affairs in nature go on in such a fast-and-loose way.

Thus far I have considered such relations of physical phenomena as have been found by experience to hold good in the whole range of physics—such relations as properly come under the domain of what is called law, and by law I mean mathematical precision, both in the antecedents and the results. With the exception of the original apparition of matter and of physical energy, there has not been found in the whole field of physics, by any investigator of any nationality, any kind of a phenomenon which is believed to be unexplainable on the basis of the knowledge of physical science we already possess. Of course, what we call explanation is merely presenting the antecedent factors of a given occurrence, both in quality and quantity, and a thing is fully explained when these are given so fully as to leave no reasonable doubt as to their sufficiency in the mind of one who is properly well acquainted with the data; but the data that enter into a given phenomenon are the very things most persons know least about; and a given explanation may be full and adequate, and yet, to some, seem to be wholly insufficient.

In these days one often hears about *specialists*—of their limited knowledge and inadequate preparation for giving a judgment in other fields than their own. So it has come to be reckoned that if a man has, by study and investigation in a given field, made himself a competent judge, so as to be considered an authority in that field, he is by so much less fitted to be heard in the settlement of some question foreign to that field; whereas some other man who is not known to have done anything in any field, may be called in for judgment, to the exclusion of the former, lest his increased knowledge in some one department should disqualify him elsewhere.

Do we not hear that biologists are incompetent judges of mental phenomena, that astronomers are not competent in biological questions, and so on? If this distinction be true to the extent generally assumed, then philosophy itself is impossible; for if a man's opinion can be good only in a small department of knowledge, and he cannot adequately master more, how shall we ever know the relationships that constitute philosophy? The truth is, this is a one-sided affair altogether, and holds true from but one standpoint. If an astronomer propounds a chemical theory of the sun, will it be needful in any degree that the chemist who reviews the work shall have even studied astronomy or paid the slightest attention to telescopes or solar affairs? If chemical science is involved, it is for the chemist to say whether what is propounded is adequate or not. That is to say, the man who concerns himself with the constitution of the sun must so far be a chemist, but a man may be a chemist and never concern himself about the sun.

Again, if a biologist who is admittedly ignorant of chemical and physical science makes statements that plainly contradict the laws of energy as determined in chemistry and physics, and if a physicist challenges the statements, shall the latter be silenced by calling him a specialist who may be competent enough in his own field, but who knows nothing of biology? Or shall he be told that physical laws may be rigorous enough in one mass of matter, but not in another? Is it to be believed that physical laws thus play fast and loose? Here the arithmetic holds good, but there all is indefinite, and would not this be a fine example of dictation out of one's field? Physiologists tell us that ultimately every physiological problem reduces itself to one of chemistry and physics.[1] If this be so, is it not plain that the one who treats broadly of biological problems must either be a physicist or submit his work to the criticism of a physicist? But a man may be a physicist and never trouble himself about biological questions.

If a social philosopher presents a scheme for ameliorating the evils present in society, in which scheme he plainly ignores the laws of life as determined by biologists,—as if such laws were not the very determining factors which must first be reckoned with,—shall not the biologist condemn such work? and shall he, too, be told that however much he knows of biology, he is incompetent in sociology? Plainly, not so. But is this process a reversible one? Can the sociologist criticise the biologist's work unless he be himself a biologist, or the biologist criticise the chemist's or physicist's work unless he be so far a chemist or physicist? He certainly cannot; and this shows that there is a certain

[1]See Appendix, p. 477.

relationship among these subjects in which there is an order of dependence. In order to fully understand and explain a sociological problem, a knowledge of psychology is essential; a working knowledge of biology, or the laws of life, and no adequate knowledge of this can be had without a preparation in chemistry and physics.

In this there is nothing new, but it is generally ignored by most persons who treat on broad questions. It is plain that every kind of a question is, in the last analysis, referable to the laws of physical phenomena, and from these there is no appeal. There are not many who like this, it is true; but the test for truth is not what one likes or dislikes, but whether the proposition is in accordance with the best and most fundamental knowledge we have. Some of those fundamental truths discovered within the past fifty years, and not questioned by any one who can stand an examination on them, were given on page 427; and whoever sees, or thinks he sees, a phenomenon which he interprets in a way which plainly contradicts or ignores those laws, does not so much have a contention with any man as with science itself. If those laws are not irrefragably true, then we have no science at all, no philosophy, knowledge is scrappy, and what we call the interdependence of phenomena is a myth.

Some of the phenomena alleged to happen at spiritual *séances*, such as levitation of human bodies, writing between closed slates, the moving of matter without contact, and so forth, are said to be as thoroughly proved as any of the facts of the fundamental knowledge I have treated. Such a statement cannot have come from any one who knows how the knowledge I spoke of was obtained, or how it may be verified by anybody who cares to take the pains. None of it depends in any degree upon anybody's dictum. If any one has doubts as to the constitution of water, he can determine it himself in half a dozen different ways. If he doubts that the earth is eight thousand miles in diameter, he can measure it in several ways. If he thinks a pound of coal does not have eleven million foot-pounds of energy, he can himself try it and be satisfied. Any one can satisfy himself by himself; assistance of others is only a convenience, not a necessity, and the fundamental statements are now believed by so many because so many have tested them, and all have reached the same conclusion. Furthermore, great commercial enterprises are founded upon some of them, as when so much limestone and coal are mixed with a given ore of iron for its reduction. So if such alleged facts be true, it cannot be true they are as thoroughly proved as the ones I stated, and they will not be so proved until each one can be verified in like manner.

There is another excellent reason for denying that they are proved in any scientific sense. All physical phenomena, so far as they have become a part of physical science, have been examined and reported upon by physicists; and both phenomena and their interpretation have been the subject of remorseless criticism, and have been adopted, if at all, on *compulsion*; their acceptance has been a matter of last resort. This is true in all departments. Why should one believe that the world turns round unless there is no other possible way to explain and account for all the facts which must be reckoned with in any explanation? The theory itself is so remote from the common experience of mankind that nobody suspected it for thousands of years, and it is not at all obvious to one who is not acquainted with phenomena out of the range of ordinary experience. The form of the earth, the aberration of light, the apparent change of latitude, and so forth, have to be considered even more than the recurrence of day and night. For most of the purposes of life it does not matter whether it turns round or not, and most men have no interest in the question further than that it accords or not with their other beliefs and feelings. But the answer to the question, "Does it turn?" is not one that can be settled by submitting it to the vote of the world. The judgment of one Galileo is worth more than that of all the rest of the world on that point. Once admit that no department of science is independent of other departments, and that no phenomenon occurs independent of relations which must be satisfied by any attempted explanation, and it follows that no explanation of an event should be adopted and be considered a part of science, unless it is shown to be in agreement with what is known. Hence, if an event is reported which appears to be out of relation with those established relations which there is general agreement upon, there is the best of reasons for thinking that either the event did not happen, or that it did not happen as reported, especially if the one reporting it is unacquainted with the variety of ways in which it is possible to do the same thing. If one sees a wheel turning round but does not see its connections, how can he tell whether it is turned by muscular action or water-power or wind-power or gravity or heat or electricity or magnetism, every one of which is capable of turning a wheel? Even if he can see the connections, he cannot always tell what makes the wheel go without further investigation. Air and steam will make a water motor go as well as water itself, and the presence of electrical devices would not insure that the wheel was turned by electricity, and the absence of such electrical devices would not insure that it was not driven by electrical agency. Hence the testimony of witnesses only, even though they were otherwise competent, would be of little weight in deciding what made the wheel go. If the question were one of any importance it could be determined only by a competent investigator with proper appliances, and unhindered by restrictions of any sort. One cannot trust his sense of sight implicitly. Many persons have lost fingers because the buzz saw looked as if it was still; and it is easy with the zoetrope, and in other ways, to produce the impression of movements that are not taking place; so it might be that after all the wheel was not turning, or even that there was no wheel at all.

Admitting, for the argument's sake, that the alleged phenomena at *séances* are real occurrences and must be accounted for, there are certainly three different possible ways:—

1. By more or less skilfully devised tricks, and fraudulent only in the attempt to make others believe they are not tricks. To be certain they are not the results of manipulative skill on the part of some one, only a skilful juggler might be able to find out. It is known that hundreds have been thus imposed upon; and skilful jugglers, such as Hermann and Maskaline, who have investigated many such, declare themselves

satisfied that the whole of it is trickery.

2. Suppose some of the surprising things done are not the results of conscious duplicity, then it may be, as most interested persons contend, the work of disembodied spirits who, through the agency of mediums, do apparently the most absurd and irrational things, but are never willing or able to do the simplest reasonable thing to satisfy a competent judge; who mutter no end of maudlin rubbish, add nothing of wisdom or knowledge to mankind, and justify Professor Huxley in saying that if such is the state of the dead we have another good reason against suicide.

3. There are a small number who think some of the *phenomena* to be genuine, but who attribute them not to spirits, but to some obscure physical force not yet understood, and but little investigated. This is the attitude of Professor Crookes, and of the Milan experimenters.

As to the class that is satisfied with the spiritistic interpretation, it may be remarked that such an explanation is in accordance with the attempts of the race to give a rational explanation of all kinds of phenomena. In the absence of proper knowledge, what seems simpler or more natural than to assume some intelligent agency as the cause of any obscure event? This it was that peopled the mountains, glens, trees, and rivers with unseen beings, watchful and interested in the affairs of men. The more ignorant, the closer was the fetich; the more enlightened, the higher these agencies retreated into the sky, useful now chiefly for literary and artistic purposes. For some reason it has always been discreditable to be without some theory for all sorts of occurrences, and even to-day, in the most enlightened communities, a man is liable to be denounced for his stupidity or his cowardice if he says, about some matters, I don't know. It is said, however, that some of the phenomena at *séances* bear the marks of intelligence such as do not belong to natural occurrences, and that it is a fair inference that other minds than the witnesses are present. When Kepler discovered that the planets moved in elliptical orbits instead of circular ones as had been supposed, he felt bound to give some reasonable explanation of the facts. He knew of nothing but intelligence that could maintain such motions, and he therefore supposed that each planet must have some guiding spirit. When the law of gravitation was applied, it was found that a circular orbit was the only unstable orbit in the system, and that gravity alone was sufficient to account for the order, the harmony, and all the variety of motions; so the spirits were dismissed from further duty. When a spider has a leg grow to replace one that has been lost, it has been held to be due to intelligent action superior to ordinary chemical and physical action. When a crystal of quartz is seen to replace a part accidentally lost, so as to complete its symmetry before it begins to grow elsewhere, it appears as if mind was at work here quite as much as in the other case, only in the latter most persons are content not to follow the implications, for they quickly see the philosophical rocks ahead. The real truth is that the further one pursues the causes of phenomena the more clearly does it appear unlikely that disembodied intelligence is behind any particular phenomena.

Among all those who make up the great class of believers in the spiritualistic theory of physical phenomena, there is not a single physicist; that is, not one to whom one would go for an explanation of any complicated physical process. It is assumed that he is no better qualified to investigate *séance* phenomena than others who do not know what to expect and look out for in simpler cases, and that he is unreasonable if he does not accept the statements of untrained observers as being as good as his own observations.

It is true that he has some prepossessions. He does not believe the multiplication table should be trifled with. He knows that most things may be done in many different ways, independent of appearances. He knows a man may sometimes not perceive what is plainly before his eyes, simply because he was not looking for it. He deems it right to exhaust the possibilities of the known before summoning some unknown and hypothetical factors in any given case. He knows it to be well-nigh impossible for a man to give an entirely accurate account to-day of what occurred yesterday. He knows that a photograph is a better witness of an event, and that a stenographic report of statements made is more reliable than any man's memory. He knows that the interpretations of events by mankind have never been true interpretations, and that the general beliefs of mankind have never been confirmed by science in any particular, and that, so far as anything has been settled, it has been decided against the opinions and judgment of mankind and its leaders. He is aware that his key has unlocked every one of the doors in Doubting Castle that have been unlocked, and therefore he believes that the implications of physical science as a whole are against any generally received interpretation of any event that has not been subjected to its scrutiny.

CHAPTER XVI

The Relations of Physical and Psychical Phenomena[1]

KNOWLEDGE has grown apace within the past fifty years. It is generally admitted that more has been acquired in this time than in all the preceding centuries. Furthermore, the knowledge thus acquired has not been simply an addition to the mental possessions of former days; it has instead been of such a kind as to completely overthrow nearly all former notions of nature and its mode of operations, and the new product can hardly be allowed to be an outcome of the work of earlier men. It is in the nature of a catastrophe where old continents have sunk and new ones have arisen from old ocean beds.

This generation lives in a new world, with new environments, new ideas, new explanations, new philosophy, new ideals, and new beliefs. We have new astronomy, new chemistry, new physics, new psychology, new natural history, and everybody is on the *qui vive* to know what can possibly come next.

[1]Read before the Psychical Congress, Chicago, August, 1893.

This does not mean that nature goes on in a different way from what it had hitherto done, but that we have mentally grasped a new and transforming idea. We have reached an elevation from which it is possible to survey a broader field, and can interpret phenomena better because their relations are better perceived, and because of this it is seen that the old interpretations were all wrong, and, indeed, were worthless, because not true. While all this is granted readily by most thoughtful persons, there are not a few who recognize the changed opinions in the various sciences and philosophy in general, who are not at all persuaded but what the present philosophy of things, which is dubbed evolution, is only a passing phase and may itself presently give way to some new and possibly truer conceptions, being content to be mildly agnostic on such matters, and willing to wait with patience for more light. There are some who think the new philosophy does not take account of all the known factors, if, by chance, there may not be unknown factors of as much or more importance than any which have been included, and which a final philosophy of things will certainly include; and such object strenuously to the limitations which the current philosophy seems to set to knowledge and to the ideals of the race.

The man of science hears rumors of phenomena

which are said to be as certain as any in his own field, which he has never investigated, and which cannot come into his category of related things. Some of these reported happenings are as marvellous as any miracles that have been recorded. Persons of undoubted probity have reported phenomena taking place in their presence which, if true, give credence to many things for which in the past men and women have been burned to death as wizards and witches. Thus, I have an acquaintance, an eminent man not given to romancing, who assures me he has seen, in undimmed light, a chair ten feet from any person rise as if some one had hold of its back and come and set itself down by his side. Something of the same kind is said to have taken place in the Milan experiments of last fall. Mr. William Crookes tells us that the weight of a body has been changed to be more or less according to an effort of the will of Mr. Home, and likewise in Milan the weight of the medium varied as much as fifty pounds.

Now, there have been numerous attempts to define a miracle for the purposes of philosophy, and usually it is not the thing accomplished so much as the means adopted for doing it. The antecedents of the event are supposed to be other than the usual ones which might do the same thing. Thus, a chair may be moved by a person who lifts it and carries it to a new place; but the chair may be pushed by a stick or pulled by a string to a new place, while no one touched it, and all who have been to see Hermann, and other magicians, have seen things move about in a surprising manner when no one touched them. In such cases it is believed that none but well-known means are skilfully used to produce such displacements, and that any one might learn the art if it were worth his while. In other words, no one thinks he is looking at a miraculous event at a magician's show, no matter how surprising the thing done; but if any person should be able to make a chair, or an object, move from one place to another without the mechanical adjuncts of some sort which are needed by others, by an act of will rather than by the employment of what we call energy, such a person is able to work what has always been called a "miracle." His method of doing that thing is a super-natural method, which is not the gift of every one even in the slightest degree; for any one can try and satisfy himself as to whether he can, by any simple act of will, make the tiniest mote in a sunbeam or the most delicately poised needle move in the slightest degree. This is the common experience; and because it has been found by experience that matter never moves except when some other body has previously acted upon it with a push or a pull, it has come about that we have reduced the experience to the statements embodied in so-called laws of motion, have found them to be justified and without any exception so far as investigation has gone, and this, too, by a multitude of persons for two hundred years. As modern science rests upon a mechanical basis, as it is concerned altogether with the phenomena of matter and the relations of the phenomena, and as these have been found in every case that has been fully investigated to conform to mathematical laws rigorously, not partly or dubiously, is it not much more probable that any other phenomenon, no matter what, that involves matter and its changes, does conform strictly to the general laws, than that these laws are sometimes inoperative?

Probably the whole thing resolves itself into this: Are the fundamental properties of matter variable? Some of the phenomena alleged to happen at *séances* imply that they are. How strong the case is against such assumption, I think is not perceived by many persons who give credence to the happenings, but who are not well equipped with physical knowledge. Many persons seem willing enough to admit physical laws and physical processes in what they take to be the field of physics, but they hold that there are other fields just as certain, and among such, mind, that controls matter and its forces, and to which it is not necessarily subject; that it is perfectly philosophical to think that mind may exist independent of matter and its relations, and be able in this condition to control phenomena.

Let us examine this. Assume that every physical process in the world should be suddenly stopped, so there should be no change. That would mean that all motions were stopped. There would at once be neither day nor night, for these are due to the earth's rotation; no light, for light is a wave motion; there would be no heat, for heat is a vibratory motion; there would be no chemical changes, for they depend upon heat; there would be neither solid nor liquid nor gas, for each depends upon conditions of temperature, that is, of heat, which is assumed to be absent; there would be no sight, for that implies wave motions; nor sound, for that implies air waves; nor taste, for that implies chemical action; nor smell, for like reason; nor touch, for that implies pressure—the result of motion. The heart would cease to beat, the blood to flow, and consciousness would be stopped. Every one of the senses would be obliterated or annihilated; nothing would happen, because there would be no change anywhere. Every phenomenon in the world of sensation would be stopped, because every phenomenon in the physical world had stopped; which is the same as saying that all we call sensations are absolutely dependent upon physical changes, going on all the time independent of our will or choice, and which cannot be controlled in the slightest degree by anybody. Every phenomenon of every kind, then, consists in, as well as is dependent upon, matter and its motion, and there is in the whole range of experience no example of any kind of a phenomenon where matter, ordinary matter, is not the conditioning factor. There is no known case where force or energy is changed in degree or direction or kind but through the agency of matter. Every kind of a change implies matter that has thus acted. What is called the correlation of forces means that one kind is convertible into some other kind of energy, as heat into mechanical energy in the steam engine. But the engine, a material structure, is essential for the change. What is called the conservation of energy means that in all the exchanges energy may undergo, as heat into light, or work of any kind, the quantity of it does not vary. The matter, as such, does not add to, or subtract from it; hence only a material body can possess energy, and a second material structure is necessary in order that the energy of the first should be changed into any other form. So it appears there must be at least two bodies before anything can possibly happen.

This all means that what we call energy is embodied only in matter, and that what we call phenomena is but the exchange of energy between different masses of matter; also that these exchanges take place with mathematical precision, else prediction would be impossible, and computation a waste of time.

Now, assume that the physical structure of an individual was kept intact, and that every atom and molecule in the body maintained its relative position after all motions had ceased. Assume, too, that the mind or soul, or whatever one chooses to call the conscious individuality, was present and capable as ever of acting upon the material structure; can a single atom be moved in the slightest degree? If any be moved, then energy has been expended, energy which must have existed elsewhere or have been created *de novo*. For conscious perception, whether sight or sound or any other, motions embodying energy are essential, as pointed out; and hence, to produce any perception, some motions would necessarily have to be initiated, and to initiate them energy from some source must be supplied. All the energy the matter had has been destroyed according to the assumption; so, if any movement has begun, it must have been created or produced from some other unthinkable condition which was not energy, in some such sense as matter is supposed to have been created, in which something is made out of nothing. The demand is for creative power. Admit for the argument's sake that it is done, and matter begins to move in any kind of a way; so far it possesses energy, physical energy as embodied in matter. Call the amount of it "A." Now, if the original condition of things was established, so far as the amount of energy was concerned, which may be called "B," then the whole amount of energy is "A plus B." It will make no difference in this sum if one supposes that the original motions and energy were not interrupted; for if, on account of mind action, any particle moves more or less than it would have done with its original supply, then something has been added to the store of energy in matter, and what is called the conservation of energy is not true.

Until all phenomena have been examined, there will be obscure happenings and things to be explained by some one who can; but it is no final explanation of anything to say, "A man did it," or "An intelligence did it." What kind of changes, that is, what kind of phenomena, the forms of energy we are now acquainted with are capable of producing no one can now limit, certainly not one who has not been to the pains to understand how the simple ones take place. I have often been told that things cannot move in certain ways, or certain things cannot be done except by intelligent action or guidance; but it may be remembered that

Kepler thought guiding spirits were needful for making the planets move in their elliptical orbits. If one must explain an obscure phenomenon, is it not wisest to explain it in accordance with what we know rather than in accordance with what we do not know? It is better for one to acknowledge his ignorance of the cause of it, than to go romancing for a reason, and repudiate all we really do know and its implications. A juggler may do the most surprising things before one's eyes, but if one cares to inquire into the antecedents of anything done he will have no difficulty in tracing it as far as the breakfast. What is meant is, the juggler does nothing which does not require energy,—energy of the ordinary sort, in the same sense as if it had been required for sawing wood or walking up the street. As for consciousness, dexterity, and all that is implied in both, I pointed out a little way back there could be neither in the absence of those changes which constitute physical phenomena; and that not only life itself, but consciousness as we know it, would be impossible without the exchanges in the energy embodied in the cellular structure of the brain. In the light of what has been accomplished in the direction of physiological psychology, it is entirely unwarrantable to assume that even thinking can go on in the absence of physical changes of measurable magnitude; and this is the same as saying that what we call intelligent action is physical at its basis.

There is such a formal agreement as well as actual connection between conscious life and the life of the brain, that it is not to be supposed any one who has properly attended to the facts will venture to deny them. Argue as one will, it is true there is no experimental knowledge that is a part of science, of consciousness separable from a material structure called brain, in which physiological changes take place as the conditions for thinking as well as for acting. This is the only known relation of mind and body. However this association of such apparently different provinces is to be explained, it is still true that for every phenomenon in consciousness there is a corresponding phenomenon in matter. Psychologists have pointed out that the phenomena indicate an identity at bottom between the activity of consciousness and cerebral activity. To follow this out into particulars would be interesting and perhaps profitable to most; but the significance of it here is that even in the psychological field, where the opportunities for investigation are right at hand and most is known, there is no evidence for consciousness apart from a material structure, or that the law of conservation of energy does not hold as strictly true here as elsewhere in physics. So there is no experimental reason for assuming the existence of incorporeal intelligences. There is no psychological question that is not at the same time a physiological question.

Experimentally it appears that the association of mind with matter and energy is not of such a nature that one is at liberty to assume their dissociation, any more than one is at liberty to assume gravitation or magnetism as independent existing somethings controlling matter according to certain laws. So any hypothesis invented to account for an occurrence that is not yet explained ought not to be in contradiction to everything else we know, and ought not to be entertained except as a last resort; and the hypothesis of disembodied intelligences acting now in and now out of the field of material things is such an one. If such phenomena really happen at *séances* as are alleged, then we have to do with affairs strictly within the line of physics, whether such phenomena are so-called mental or so-called physical. It is useless to affirm that the two are such radically different phenomena that the methods of the latter are not appropriate in the former; and the extensive laboratories for physiological psychology, which are now being established in all the larger institutions of learning, is a sufficient denial of the proposition.

The term psychics is intended to denote something

different from the phenomena of psychology as manifested in a given organism. It is supposed to relate to the sympathetic relation of one mind to that of another quite apart from the ordinary physical relations, that is, from the senses. As for the mind-reading as exhibited some years ago by Brown and others, I believe it is now agreed that it is due to the sense of touch, and cannot be done without contact. In hypnotic work there has to be "suggestion," and most of the very remarkable cases, such as those in France last winter, have been shown to be gross frauds. But let it be granted that some of it is genuine, that it is possible in some cases to impart information and discover the thoughts of another without the common resources, it does not then follow that the method is extra-physical. If only here and there is to be found an individual called a psychic, who is thus sensitive, and it is not a race endowment, one no more need to summon a mysterious supernormal agency to account for it, than such is needed for the work of Newton or Mozart. Because a phenomenon has not been explained, and no one knows how to explain it, is no reason at all for supposing there is anything mysterious about it. There are any number of phenomena throughout nature that have not been explained, and no one knows how to explain on the basis of what is known. Such, for instance, is the whirlwind that crosses the field, raising dust and leaves into the air. No one has explained the soaring of birds, no one knows what goes on in an active nerve, or why atoms are selective in their associates. Ignorance is not a proper basis for speculation; and if one must have a theory, let it be one having some obvious continuity with our best physical knowledge.

What is here given is not intended to be a denial that such phenomena as thought-transference, or even the most surprising things such as those described in the Milan experiments, take place. It is only intended to emphasize the probability that whatever happens has a physical basis, and is therefore explained only when these physical relations are known.

APPENDIX

NOTE TO PAGE 67.

As to whether it is considered as known that the sum of the interior angles of a plane triangle are exactly equal to one hundred and eighty degrees: "Suppose that three points are taken in space, distant from one another as far as the sun is from α Centauri; and that the shortest distance between these points is drawn so as to form a triangle. And suppose the angles of this triangle to be very accurately measured and added together; this can at present be done so accurately that the error shall certainly be less than one minute, less therefore than the five-thousandth part of a right angle. Then I do not know that this sum would differ at all from two right angles; but I also *do not know that the difference would be less than ten degrees,* and I have reasons for not knowing."

W. K. CLIFFORD: *Aims and Instruments of Scientific Thought.*

"If the Euclidian assumptions are true, the constitution of parts of space at an infinite distance is as well known as the geometry of any portion of this room.

So that here we have real knowledge of something at least that concerns the cosmos; something that is true throughout the immensities and the eternities. That something Lobotchewski and his successors have taken away."

W. K. Clifford: *Philosophy of the Pure Sciences.*

"In this case the universe as known becomes a valid conception, for the extent of space is a finite number of cubic miles. If you were to start in any direction whatever, and move in a perfectly straight line according to the definition of Liebnitz, after travelling a most prodigious distance . . . you would arrive at—this place."

Ibid.

"It must remain an open question whether, if we had large enough triangles, the sum of the three angles would still be two right angles."

Enc. Brit. 9th ed., Art. Measurement.

"It is true that according to the axioms of geometry, the sum of the three angles of a triangle are precisely one hundred and eighty degrees; but these axioms are now exploded, and geometers confess that they, as geometers, know not the slightest reason for supposing them to be precisely true. That they are exactly that amount is what nobody can be justified in concluding."

C. S. PEIRCE: *Monist*, vol. i. No. 2, p. 174.

"All that we need do is to call the attention of those who busy themselves with mental philosophy to this generalization of geometry as one of the results of modern mathematical research which they cannot afford to overlook."

GEORGE CHRYSTAL, *in Enc. Brit., Art. Parallels.*

Such as care to look into the matter further will find the subject treated in an untechnical way in the works of W. K. Clifford, in the chapters on the "Theories of the Physical Forces," "Aims and Instruments of Scientific Thought," and especially the "Philosophy of the Pure Sciences." There is much on it in the *American Journal of Mathematics*, vols. i. and ii., also in the "Proceedings of the Royal Society," Edinburgh, vol. x., 1879, and in article "Measurement," *Enc. Brit.*

NOTE TO PAGE 249.

In 1881 the author discovered how electric ether waves could be produced and identified, where the vibratory rates were as high as 4000 or more per second, by employing static telephones detached and removed many feet from the inducing electric current. These gave a wave length of $\frac{186000}{4000} = 46+$ miles long. Hertz, Tesla, and others have since then described methods of producing them so short as to be but a few feet long. When they have thus been mechanically shortened so as to be but the one forty-thousandth of an inch in length, they will be seen by the eye as red light.

NOTE TO PAGE 291.

See Maxwell's "Theory of Heat," pp. 160, 161.

$\frac{H}{S} = \frac{h}{T}$. S and T are the absolute temperatures of the hot and cold bodies in Carnot's engine. H and h are the quantities of heat taken up and given out. When $T = 0°$, $h = 0$, when h is the equivalent of the work done. As this is 0 at absolute zero, no work could be done in changing the volume of a substance at that temperature. There can be no cohesion among the molecules or atoms, for this would require that work should be done to separate them. It is the temperature of *dissociation*.

This conclusion is one to which chemists and physicists have been led by their researches. For example, Dr. Lothar Meyer says, "At the lowest temperature to which we can attain, the majority of chemical reactions studied under these conditions have been found to cease or to proceed very slowly, so that it would appear to be very probable that at the absolute zero, viz., 273°, a temperature much below the lowest yet attained, chemical action would cease altogether from the absence of any form of heat motion whatsoever; so without heat there would be no exertion of the so-called chemical affinity."

Modern Theories of Chemistry, § 211.

NOTE TO PAGE 336.

The hypothesis of a "vital principle" is now as completely discarded as the hypothesis of phlogiston in chemistry. No biologist with a reputation to lose would for a moment think of defending it.

JOHN FISKE: *Cosmic Philosophy*, vol. i. p. 422.

"We can demonstrate the infinitely manifold and complicated physical and chemical properties of the albuminous bodies to be the real cause of organic or vital phenomena."

HAECKEL: *History of Creation*, vol. i. p. 330.

"The aim of modern physiology is to conceive all organic processes as physical or chemical."

HÖFFDING: *Outlines of Psychology*, p. 57.

"Physiologists must expect to meet with an unconditional conformity to law of the forces of nature in their inquiries respecting the vital processes. They will have to apply themselves to the investigation of the physical and chemical processes going on within the organism."

HELMHOLTZ: *Scientific Lectures*, p. 384.

"A vital element, i.e., an element peculiar to organisms, no more exists than does a vital force working independently of natural and material processes."

CLAUS & SEDGWICK: *Zoölogy*, part i. p. 10.

"In Physiology the word life is understood to mean the chemical and physical activities of the parts of which the organism consists."

B. SANDERSON: *Nature*, vol. xlviii., p. 613.

"Modern physiology interprets the phenomena of organic life by means of physical and chemical laws. An appeal to 'vital force' or to the intervention of mind, it does not recognize as an explanation of an organic phenomenon."

HÖFFDING: *Outlines of Psychology*, p. 10.

"Physiology thus appears as a branch of applied physics, its problems being a reduction of vital phenomena to general physical laws and thus ultimately to the fundamental laws of mechanics."

WUNDT: *Lehrbuch der Physiologie*, p. 2.

"It must not be supposed that the differences between living and not living matter are such as to justify the assumption that the forces at work in the one are different from those to be met with in the other."

Huxley: *Art. Biology, Enc. Brit.*, p. 681.

"Zoölogy, the science which seeks to arrange and discuss the phenomena of animal life and form as the outcome of the operation of the laws of physics and chemistry."

Lankester: *Art. Zoölogy, Enc. Brit.*, p. 803.

"If corporal functions are mediated by immaterial agencies, physiological science is impossible."

G. Stanley Hall: *Amer. Jour. Psychology*, vol. iii. p. 74.

"It has not occurred to me that any one now uses the term 'vital force' in any other way than as a convenient method of expressing the sum total of the physical and chemical activities of organisms."

Prof. E. L. Mark, *Harvard University.*

"These phenomena of life, though they may not as yet be physically and chemically explained, are certainly not to be referred to the working of any special *vital force* peculiar to organisms.... We have to do here with the same forces and the same substances that we meet with elsewhere in nature."

Lang: *Textbook of Comp. Anat.*, London, 1891, p. 2.

"Modern science has allowed the vitalistic theory

(*vitalismus*) to drop; instead of by means of a special vital force, it explains irritability as a very complex chemico-physical phenomenon. It is only distinguished from other chemico-physical phenomena of inorganic nature by degree, namely, that the external stimuli come in contact with a substance of complicated structure, an organism, and correspondingly produce in it also a series of complicated processes."

O. Hertwig: *Die Zelle und die Gewebe*, p. 75, 1893.

"I know of no authority in recent years which recognizes a distinct vital force; all students of nature, so far as I am aware, explain all the phenomena of life by means of physical and chemical forces."

Prof. J. S. Kingsley, *Tufts College.*

INDEX